Fitting the task to the Man

Etienne Grandjean

Former Director of the Department of Ergonomics and
Hygiene, Swiss Federal Institute of Technology, Zürich

Fitting the task to the Man

A textbook of Occupational Ergonomics

4th Edition

Taylor & Francis
London • New York • Philadelphia
1988

UK	Taylor & Francis Ltd, 4 John Street, London WC1N 2ET
USA	Taylor & Francis Inc., 1900 Frost Road, Suite 101, Bristol, PA 19007

Originally published 1963 by Ott Verlag Thun under the title
Physiologische Arbeitgestaltung
2nd Edition published 1969
3rd Edition published 1979 by Ott Verlag Thun. All rights reserved.
© 1979 Ott Verlag Thun
English edition translated by Harold Oldroyd and published 1980 by
Taylor & Francis Ltd, 4 John Street, London WC1N 2ET
© 1980 Taylor & Francis Ltd

4th edition copyright © E. Grandjean 1988
Reprinted 1990, 1991

British Library Cataloguing in Publication Data
Grandjean, Etienne
    Fitting the task to the Man—4th ed.
    1. Human engineering
    I. Title
    620.8'2   TA166
    ISBN 0-85066-380-6
    ISBN 0-85066-379-2 Pbk

Library of Congress Cataloging-in-Publication Data
Grandjean, E. (Etienne)
    Fitting the task to the man.
    Translation of: Physiologische Arbeitsgestaltung.
    Bibliography: p.
    Includes index.
    1. Work—Physiological aspects.   2. Human mechanics.
3. Human engineering.   4. Industrial hygiene.   I. Title.
QP309.G7313   1988   612'.042   87-33573
ISBN 0-85066-380-6
ISBN 0-85066-379-2 (pbk.)

Cover design by Russell Beach

Typeset by Mathematical Composition Setters Ltd, Salisbury.

*Printed in Great Britain by Burgess Science Press,
Basingstoke, Hants*

Contents

About the author

Professor Etienne Grandjean has been one of the leading figures in Ergonomics in Europe for over 30 years. Born in 1914 in Berne, Switzerland, he obtained his MD in 1939 and became Director of the Department of Hygiene and Ergonomics at the Swiss Federal Institute of Technology, Zurich in 1950, where he remained until his retirement in 1983. His main research interests were the sitting posture, fatigue and working conditions in industries, and for the last decade VDT workstations.

Professor Grandjean was granted several international awards and received Honorary Doctorates from the Universities of Surrey, Stuttgart and Geneva.

He was one of the founders of the International Ergonomics Association and its General Secretary from 1961 to 1970. He has published some 300 scientific papers and edited the first edition of *Fitting the task to the Man* in 1963, which has since been translated into 10 different languages. He has also written two other books in English — *Ergonomics of the Home* (1973) and *Ergonomics of Computerized Offices* (1987).

Preface to the fourth edition

The word ergonomics derives from the Greek *ergon*, meaning work, and *nomos*, which is best translated as natural law or system.

In the first forty years of its existence ergonomics has undergone a considerable diversification; it broke away from the industrial area and took possession of new fields, namely that of consumer products, the home, road traffic and safety, hospitals and schools as well as sport and leisure. But the primary aim of ergonomics has remained the same — to optimize the functioning of a system by adapting it to human capacities and needs.

A book on ergonomics today can hardly cover all its disciplines and applications. Thus the present edition is, as with the preceding ones, concentrating on basic issues of ergonomics and their occupational applications and endeavours to offer the reader as many practical hints as possible.

The last edition was prepared in 1978. Since then we have gone through the exciting period of the development of information technology with the overall penetration of computer displays mainly into offices. Thus white-collar workers today are tied to man—machine systems. They have become susceptible to constrained postures, to inadequate lighting conditions and to visual strain. This is the reason why in our time offices have become an important part of occupational ergonomics. It is logical, therefore, to devote several chapters of the new edition to problems of visual display terminals and their environment.

Most chapters of the previous edition have been shortened in favour of new topics, among them the design of work stations and keyboards, recent progresses towards the thorny problems of back troubles, occupational stress and job design as well as the physiology of reading and visual strain.

The aim and purpose of this new edition remains the same — to impart the important elements of ergonomics in a simple and lucid form to all those who are responsible for putting into practice the principles of occupational ergonomics.

Zurich   *Etienne Grandjean*
15th December, 1987

Acknowledgment

My thanks are due to Mrs Ilse New-Fannenböck for correcting my English and rendering it 'user-friendly'.

1. *Muscular work*

1.1. Physiological principles

Structure of a muscle

The human body is able to move because it has a widely distributed system of muscles, which together make up approximately 40% of the total body weight. Each muscle is made up of a large number of muscle fibres, which can be between 5 mm and 140 mm long, according to the size of the muscle. The diameter of a muscle fibre oscillates about 0·1 mm. A muscle contains between 100 000 and 1 million such fibres, each of which is drawn out into a sinew at each end. The fibres of long muscles are sometimes bound together in bundles. At each end of the muscle the sinews are combined into a tough, non-elastic tendon, which in turn is firmly attached to the bony skeleton.

Muscular contraction

The most important characteristic of a muscle is its ability to shrink to half its normal length, a phenomenon we call muscular contraction. The work done by a muscle in such a complete contraction increases with its length: for this reason athletes try to make their muscles longer by stretching exercises.

Each muscle fibre contains proteins, among which actin and myosin assume special importance, since this system is contractile, and contributes most to the muscular contraction. Actin and myosin are present as fibres which can slide over each other. During the process of contraction the actin fibres are believed to ensheath themselves in between the myosin fibres, as illustrated in Figure 1.

Muscle power

Each muscle fibre contracts with a certain force, and the strength of the whole muscle is the sum of these muscle fibres. The maximum strength of a human muscle lies *between 0·3 and 0·4 N/mm^2 of the cross-section* [112]: thus a muscle of 100 mm^2 cross-section can support a weight of 3−4 kg (30−40 N). Hence a person's inherent muscular strength depends, in the first instance, on the cross-section of his or her muscles. Given equal training, a woman, with her narrower muscles, can exert on average about 30% less force than a man

1

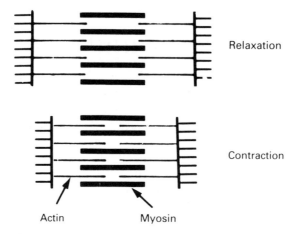

Relaxation

Contraction

Actin   Myosin

Figure 1. Model of muscular contraction.
The actin fibres slide between the myosin fibres and the two ends of the muscle are drawn closer together.

(Scherrer, [231]). A muscle produces its greatest force at the beginning of its contraction, when it is still at its relaxed length. As the muscle shortens, its power declines. Many of the recommendations in Chapter 3 for improving working efficiency are based on this relationship.

Regulation of muscular effort

The number of actively contracting muscle fibres determines how power is developed during the period of contraction. As we shall see later, a muscle fibre is made to contract by incoming nervous impulses; hence the amount of muscle power produced is determined by the number of nervous impulses, that is by the number of motor nerve cells in the brain that have been excited. The speed of a muscular contraction depends upon how quickly power is developed during a given interval of time, so the rapidity of a movement is governed by the number of actively contracting muscle fibres.

When muscular contraction is slow, or maintained for a long time (static muscular work), muscle fibres are brought into active contraction in succession, alternately. Each individual fibre has resting periods, which permit a certain amount of 'recuperation' to take place.

Sources of energy

During contraction, mechanical energy is developed at the expense of the reserves of chemical energy in the muscle. Muscular work involves the transformation of chemical into mechanical energy. The energy released by chemical reaction acts on the protein molecules of the actin and myosin fibres, causing them to change position, and so bring about contraction. The immediate sources of energy for contraction are energy-rich phosphate compounds which change from a high-energy to a low-energy state in the course of chemical reactions. The source of energy most widely used by living organisms is *adenosine triphosphate* (ATP), which releases considerable amounts of energy when it is broken down into adenosine diphosphate. Moreover, ATP is present not only in muscles, but in nearly every kind of tissue, where it acts as a reservoir of readily available energy. Another source of chemical energy in muscle fibres is phosphocreatine (phosphagen) which releases an equally significant amount of energy when it is broken down into phosphoric acid and creatine.

The low-energy phosphate compounds are continuously converted back to the high-energy state in the muscles, so that the reserves of energy remain undiminished. This is one of the wonders of nature: it is as if the exhaust gases of a car could be re-converted into petrol!

The role of glucose, fat and proteins

This regeneration of high-energy phosphate compounds itself consumes energy, which is obtained from glucose and from components of fat and proteins. Glucose, the most important of the sugars circulating in the blood, is the main energy supply in intensive physical work. Under conditions of rest or moderate physical work the components of fat (fatty acids) and of proteins (amino-acids) are the dominant energy supplies. *These nutritive substances, glucose, fat and protein, are, therefore, the indirect energy sources for the continuous replenishment of energy reserves in the form of ATP or other energy-rich phosphate compounds.* The glucose passes out of the bloodstream into the cells, where it is converted by various stages into *pyruvic (pyro-racemic) acid*. Further breakdown can take two directions, depending on whether oxygen is available (aerobic glycolysis), or whether the oxygen supply is deficient (anaerobic glycolysis).

The role of the oxygen

If oxygen is present, then the pyruvic acid is further broken down by oxidation (i.e., under continuous oxygen consumption), the end-products being water and carbon dioxide. This releases enough energy to reconstitute a large amount of ATP.

If oxygen is lacking, then the normal breakdown of pyruvic acid does not take place. Instead it is converted into *lactic acid*, a form of metabolic waste product which plays a vital part in symptoms of muscle fatigue and 'muscular hangover'. This process releases a lesser amount of energy for the reconstitution of energy-rich phosphate compounds, but it allows a higher muscular performance under low-oxygen conditions, at least for a short time.

Oxygen debt

After heavy muscular effort a person is said to be 'out of breath'. This means that he is making up for a shortage of oxygen by breathing more heavily; *he is paying off his oxygen debt*. This oxygen debt arises from previous consumption of energy; the extra oxygen is needed to convert lactic acid back

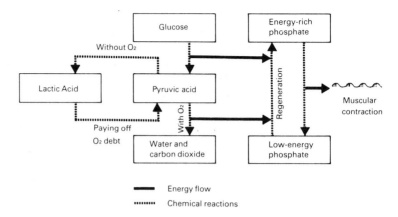

Figure 2. Diagram of the metabolic processes which take place during muscular work.

into pyruvic acid, and to reconstitute energy-rich phosphate compounds. After that, energy can again be obtained by oxidative breakdown of pyruvic acid.

Figure 2 shows a much simplified diagram of the energy supply of a muscle.

Protein and fat   As already mentioned, fats and proteins are also involved in these metabolic changes. When breakdown of these substances has reached a certain stage, a *common metabolic pool* comes into being. The remaining fragments of fatty acids (from breakdown of fats) and amino-acids (from breakdown of proteins) undergo the same final breakdown as the pyruvic acid, and end up as water and carbon dioxide. This last stage therefore makes its own contribution of energy to the muscular effort.

The blood supply   These substances that are so important for energy production —glucose and oxygen—are only stored in small amounts in the muscles themselves. Both of them must therefore be continuously transported to the muscles by the blood. For this reason, in the final analysis it is the blood supply that is the limiting factor in the efficiency of the muscular machinery. During effort, a muscle increases its need for blood several fold, and to supply this the most important adaptations of the blood system are more active pumping by the heart, raising of the blood pressure, and enlargement of the blood vessels that lead to the muscles.

According to Scherrer [231] the following increases in circulation can be expected:

Muscle at rest	4 ml/min/100 g muscle
Moderate work	80 ml/min/100 g muscle
Heavy work	150 ml/min/100 g muscle
After a restriction of the blood circulation	50−100 ml/min/100 g muscle

Heat production   According to the first law of thermodynamics, a muscle must
be supplied with the same amount of energy that it uses. In
practice the incoming energy is transformed into: (*a*) work
performed; (*b*) heat; and (*c*) energy-rich chemical compounds.

Energy stored in the form of phosphate compounds is the
smallest component; in contrast, generation of heat is by far
the biggest. If heat-generation within the muscles is measured
with delicate instruments, the following constituent parts are
recognizable:

1. *Resting heat production* amounts to $1 \cdot 3$ kJ/min for a man
   weighing 70 kg. According to Scherrer [231] this serves
   to maintain the molecular structure and the electrical
   potentials in the muscle fibre.
2. *Initial heat* is considerably greater than the resting heat.
   This is the heat produced during the whole course of
   contraction of the muscle, and is proportional to the work
   done.
3. *Recovery heat* is produced after the contraction is finished
   and over a longer period, up to 30 min. This is obviously
   the output from the oxidative processes of the recovery
   phase, and is of about the same order of magnitude as the
   initial heat.

Associated
electrical
phenomena

It has been known for a long time that muscular contraction is
accompanied by electrical phenomena, which are very similar
to the processes of transmission of impulses along a nerve. In
recent decades these electrical processes have been studied
in more detail, using very delicate electrophysiological
methods.†

In simple terms we can make the following points:

1. The resting muscle fibre exhibits an electrical
   potential—the so-called *resting membrane potential*—of
   about 90 mV. The interior of the fibre is negatively charged
   in relation to the exterior.
2. The beginning of contraction is associated with a collapse
   of the resting potential, and an overriding positive charge
   in the interior. This reversal of potential is called the *action*
   potential, so-called because it arises 'during the action' of
   the nerve. The action potential in the muscle lasts about
   $2-4$ ms, and spreads along the muscle fibre at a speed of
   about 5 m/s.
3. The action potential involves depolarization and repolarization of the membrane of the muscle fibre. During this
   period the muscle fibre is no longer capable of being
   excited; this period is called the absolute refractory period,

---

† Several details of such electrical processes will be dealt with in Chapter 2
'Nervous control of movements'.

and lasts 1−3 ms. By analogy with the processes that go on in nerves, in the muscle fibre de- and repolarization are also manifestations of reciprocal streams of potassium and sodium ions through the membrane of the muscle fibre.

*Electro-
myography*

The electrical activity of a muscle can be recorded with the help of an amplifier, a technique known as *electromyography.*

The electric current can be tapped by attaching suitable electrodes to the surface of the skin immediately over the muscles to be studied. Another technique, less often used, is to insert needle-electrodes into muscles and thereby be able to monitor individual muscle fibres. An electromyogram registering the output of surface electrodes records the total electrical activity of the entire muscle. For this purpose two electrodes, each about 100 mm^2 in area are applied a few centimetres apart. The output of the electrodes is usually integrated and amplified electronically. Results by this method have so far had only a relative significance because they are valid only for one particular set of electrodes in one particular experiment.

Nevertheless, electromyography has shown that electrical activity increases with the level of muscular force developed.

Electromyography is specially useful for investigating the amount of muscular effort expended in different bodily attitudes.

1.2. Static muscular effort

There are two kinds of muscular effort: (*a*) *dynamic (rhythmic) effort*; and (*b*) *static (postural) effort.*

*Dynamic and
static forms of
work*

Figure 3 illustrates these two kinds of muscular activity. The dynamic example is cranking a wheel, and the static example is supporting a weight at arm's length.

The two forms of muscular effort can be described as follows:

1. *Dynamic effort is characterized by a rhythmic alternation of contraction and extension, tension and relaxation.*
2. *Static effort, in contrast, is characterized by a prolonged state of contraction of the muscles, which usually implies a postural stance.*

In a dynamic situation the effort can be expressed as the product of the shortening of the muscle and the force developed (work=weight×height it is raised). During static effort the muscle is not allowed to extend, but remains in a state of heightened tension, with force exerted over an extended period. During static effort no useful work is externally visible, nor can this be defined by a formula such as weight×distance. It rather resembles an electromagnet,

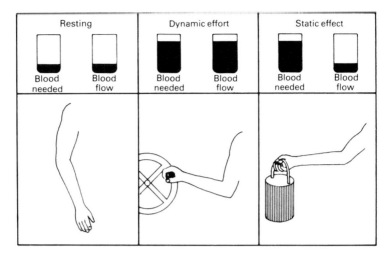

Resting		Dynamic effort		Static effect	
Blood needed	Blood flow	Blood needed	Blood flow	Blood needed	Blood flow

Figure 3. **Diagram of dynamic and static muscular effort.**

which has a steady consumption of energy while it is supporting a given weight, but does not appear to be doing useful work.

Blood supply   There are certain basic differences between static and dynamic muscular effort.

During static effort the blood vessels are compressed by the internal pressure of the muscle tissue, so that blood no longer flows through the muscle. During dynamic effort, on the other hand, as when walking, the muscle acts as a pump in the blood system. Compression squeezes blood out of the muscle, and the subsequent relaxation releases a fresh flow of blood into it. By this means the blood supply becomes several times greater than normal: in fact the muscle may receive 10−20 times as much blood as when it is resting. A muscle performing dynamic work is therefore flushed out with blood and retains the energy-rich sugar and oxygen contained in it, while at the same time waste products are removed.

In contrast, a muscle that is performing heavy static work is receiving no sugar or oxygen from the blood, and must depend upon its own reserves. Moreover—and this is by far the most serious disadvantage—waste products are not being excreted. Quite the reverse: these waste products are accumulating and produce the acute pain of muscular fatigue.

For this reason we cannot continue a static muscular effort for very long; the pain will compel us to desist. On the other hand a dynamic effort can be carried on for a very long time without fatigue, provided that we choose a suitable rhythm for it. There is one muscle that is able to work dynamically throughout our lives without interruption, and without tiring: the muscle of the heart. Figure 4 shows how the two kinds of muscular effort affect the blood supply to the working muscle.

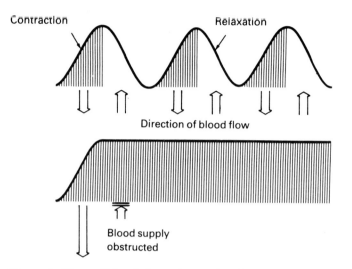

Direction of blood flow

Blood supply
obstructed

Figure 4. Flow of blood through muscles during dynamic and static effort.
The curves show the variation of muscular tension (internal pressure).
Upper: dynamic effort operates like a pumping action, which ensures a flow of blood through the muscle.
Lower: static effort obstructs the flow of blood.

Examples of static effort

Our bodies must often perform static effort during everyday life. Thus, when we stand up a whole series of muscle groups in the legs, hips, back and neck are stressed for long periods. It is thanks to these static efforts that we can hold selected parts of our bodies in any desired attitude. When we sit down, the static effort of the legs is relieved, and the total muscular strain of the body is reduced. When we lie down, nearly all static muscular effort is avoided; that is why a recumbent posture is the most restful. There is no sharp line between dynamic and static effort. Often one particular task is partly static and partly dynamic. Since static effort is much more arduous than dynamic, the static component of mixed effort assumes the greater importance.

In general terms, static effort can be said to be considerable under the following circumstances:

1. If a high level of effort is maintained for 10 seconds or more.
2. If moderate effort persists for 1 minute or more.
3. If slight effort (about one third of maximum force) lasts for 4 minutes or more.

There is a static component in nearly every form of factory work or any other occupation. The following are some of the commonest examples:

1. Jobs which involve bending the back either forwards or sideways.

2. Holding things in the arms.
3. Manipulations which require the arms to be held out horizontally.
4. Putting the weight on one leg while the other works a pedal.
5. Standing in one place for long periods.
6. Pushing and pulling heavy objects.
7. Tilting the head strongly forwards or backwards.
8. Raising the shoulders for long periods.

Constrained postures are certainly the most frequent form of static muscular work. The main cause of constrained postures is carrying the trunk, head or limbs in unnatural positions. Figure 5 shows examples of static loads.

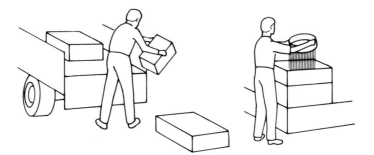

Figure 5. Examples of static muscular effort.
Left: loading parcels.
Right: sieving sand into a mould in a foundry. In both examples there are high static loads on the muscles of the back, shoulders and arms.

Effects of static work

During static effort the flow of blood is constricted in proportion to the force exerted. If the effort is 60% of the maximum, the flow is almost completely interrupted, but a certain amount of circulation of blood is possible during lesser efforts, because the tension in the muscles is less. When the effort is less than 15−20% of the maximum, blood flow should be normal.

Obviously, therefore, the onset of muscular fatigue from static effort will be more rapid the greater the force exerted, i.e., the greater the muscular tension. This can be expressed in terms of the relation between the maximal duration of a muscular contraction and the force expended. This was systematically studied by Monod [187] and later confirmed by Rohmert [217]. Figure 6 shows the results obtained by Monod for four muscles.

It appears from this work that a static effort which exerts 50% of maximum force can last no more than 1 minute, whereas, if the force expended is less than 20% of maximum, the muscular contraction can continue for some time. But field studies as well as general experience show that a static force of

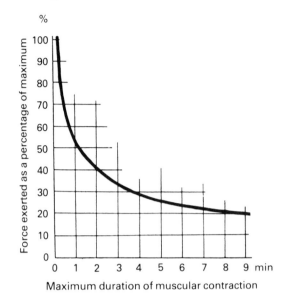

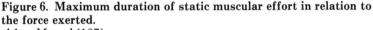

Figure 6. Maximum duration of static muscular effort in relation to the force exerted.
After Monod [187].

15–20% of the maximum force will induce painful fatigue if such loads have to be kept up for days and months [199; 271]. Many experts are therefore of the opinion that work can be maintained for several hours per day without symptoms of fatigue if the force exerted does not exceed about 8% of the maximum force of the muscle involved.

Static muscular effort is strenuous

Under roughly similar conditions a static muscular effort, compared with dynamic work, leads to:

1. A higher energy consumption.
2. Raised heart rate.
3. Longer rest periods needed.

This is easy to understand if we bear in mind that, on the one hand, the metabolism of sugar with an inadequate supply of oxygen releases less energy for regeneration of energy-rich phosphates, and on the other hand produces a large amount of lactic acid, which interferes with muscular effort. Oxygen deficiency, which is unavoidable during static muscular effort, inevitably lowers the effective working level of the muscle.

Figure 7 shows a good example of this, repeating the research results of Malhotra and Sengupta [176]. These authors showed that schoolchildren who carried their satchel in one hand needed more than twice as much energy as when they carried the satchel on their back. This increased energy consumption must be attributed to the high static loads on the arms, shoulders and trunk. Figure 8 shows another example

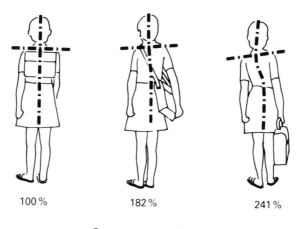

100 % 182 % 241 %

Oxygen consumption

Figure 7. Effect of static effort on energy consumption (measured by oxygen consumption) for three ways of carrying a school satchel. *After Malhotra and Sengupta [176].*

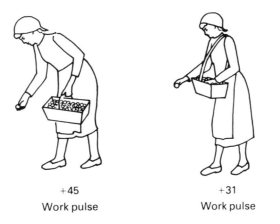

+45 +31

Work pulse Work pulse

Figure 8. Static muscular effort in the left arm during potato-planting.
The use of a sling has avoided static effort in the left arm. Over a period of 30 minutes the heart rate rose by up to 45 pulses per minute (left) or 31 pulses per minute (right). After Hettinger [113].

from Hettinger [113] involving potato planting. In one case the basket of potatoes was carried in one hand; in the other the basket hung in a harness in front of the body. Carrying the basket by hand raised the heart rate by 45 pulses per minute, compared with only 31 pulses per minute when the basket was carried in a harness. Carrying the basket required a muscular effort in the left arm amounting to 38% of its maximum. Hettinger drew the conclusion that the increased strain on the heart was entirely caused by the heavy static effort needed to carry the basket in one hand.

Combination of dynamic and static efforts

In many cases no sharp distinction between dynamic and static effort can be made. A particular task can be partly static and partly dynamic. Keyboard operating is an example for a combination of both types of muscular work: shoulders and arms do mainly static work when holding the hands in the typing position, while the fingers perform mainly dynamic work when operating the keys. Since static effort is much more strenuous than dynamic work, the static component of combined effort assumes greater importance. There is a static component in almost every form of physical work.

Localized fatigue and musculo-skeletal troubles

As already explained, even moderate static work might produce troublesome localized fatigue in the muscles involved, which can build up to intolerable pain. If the static load is repeated daily over a long period, more or less permanent aches will appear in the limbs and may involve not only the muscles but also the joints, tendons and other tissues. Thus long-lasting and daily repeated static efforts can thereby lead to a damage of joints, ligaments and tendons. All these acute and chronic impairments are usually summarized under the term 'musculoskeletal disorders'.

Several field studies as well as general experience have shown that such static loads are associated with a higher risk of:

1. Arthritis of the joints due to mechanical stress.
2. Inflammation of the tendon sheaths (tendinitis or peri-tendinitis).
3. Inflammation of the attachment-points of tendons.
4. Symptoms of arthrosis (chronic degeneration of the joints).
5. Painful muscle spasms.
6. Intervertebral disc troubles.

Persistent musculoskeletal troubles

These symptoms of overstress can be divided into two groups: *reversible and persistent musculoskeletal troubles.*

The *reversible symptoms* are short-lived. The pains are mostly localized to the muscles and tendons, and disappear as soon as the static load is relieved. *These troubles are the pains of weariness.*

Persistent troubles are also localized to strained muscles and tendons, but they affect the joints and adjacent tissues as well. The pains do not disappear when the work stops, but continue. *These persistent pains are attributable to inflammatory and degenerative processes in the overloaded tissues.* Elderly employees are more prone to such persistent troubles. According to van Wely [271], persistent musculoskeletal troubles are commonly observed among operators who work all the year round at the same machine at which the manual controls are either too high or too low.

Persistent musculoskeletal troubles, if supported over years, may get worse and lead to chronic inflammations of

Table 1. Static load and bodily pains

Work posture	Possible consequences affecting
Standing in one place	Feet and legs; possibly varicose veins
Sitting erect without back support	Extensor muscles of the back
Seat too high	Knee; calf of leg; foot
Seat too low	Shoulders and neck
Trunk curved forward, when sitting or standing	Lumbar region; deterioration of intervertebral discs
Arm outstretched, sideways, forwards or upwards	Shoulders and upper arm; possibly periarthritis of shoulders
Head excessively inclined backwards or forwards	Neck; deterioration of intervertebral discs
Unnatural grasp of hand grip or tools	Forearm; possibly inflammation of tendons

tendon sheaths or even deformations of joints. The troubles that may be expected to follow from certain forms of static load are set out in Table 1.

Examples of morbid symptoms

During a training course lasting 12 weeks, Tichauer [258] compared the effects of using a wire cutter that was shaped to the hand with those of a normal pair of pliers. The normal pliers called for an unnatural grip, with the hand rotated around the ulna, and hence a pronounced static muscular strain. During those 12 weeks, 25 of the 40 workers who used the unnatural grip showed morbid symptoms, in the sense of inflammation of the tendon attachments or of the tendon sheaths. In contrast, there were only four cases of inflamed tendon sheaths among an equal number of workers who used the curved wire-cutters which permitted a natural grip. The two forms of pliers are shown in Figure 9.

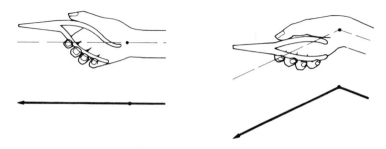

Figure 9. Left: the curved pliers are shaped to fit the hand, which remains in line with the forearm.
Right: pliers of the traditional shape require sustained static effort, with the wrist turned outwards; the hand is no longer in line with the forearm.
Modified from Tichauer [258].

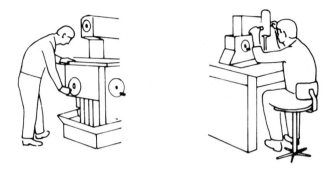

Figure 10. Unnatural postures at work which involve static loads can lead to physical disabilities.
Left: risk of back troubles.
Right: risk of aches in shoulders and upper arms.

Figure 10 shows two workplaces at a machine which could lead to a danger of aches and other unhealthy symptoms.

Standing in one place

Standing in one place involves static effort, throug.. prolonged immobility of the joints of the feet, knees and hips. The force involved is not great, and falls below the critical level of 15% of the maximum. All the same, standing in one place for a long time is wearisome and painful. This is not entirely the effect of static muscular effort; pain is also partly caused by increased hydrostatic pressure of the blood in the veins of the legs and general restriction of lymphal circulation in the lower extremities. In practice the hydrostatic pressure in the veins, when standing motionless, is increased as follows:

1. At the level of the feet by 80 mm Hg.
2. At the level of the thighs by 40 mm Hg.

During walking the muscles of the legs act as a pump, which compensates for the hydrostatic pressure of the veins by actively propelling blood back towards the heart.

Thus prolonged standing in one place causes not only fatigue of the muscles which are under static load, but also discomfort which is attributable to insufficient return flow of the venous blood.

This unhealthy state of the blood circulation is the cause of cumulative ailments of the lower extremities in occupations which call for prolonged standing without movement. Such occupations involve increased liability to:

1. Dilation of the veins of the legs (varicose veins).
2. Swelling of the tissues in the calves and feet (oedema of the ankle).
3. Ulceration of the oedematous skin.

2. Nervous control of movements

2.1. Physiological principles

Structure of the nervous system

The central nervous system consists of the brain and the spinal cord. The peripheral nerves either run outwards from the spinal cord to end in the muscles (motor nerves), or inwards from the skin, the muscles, or the sense organs to the spinal cord or the brain (sensory nerves). The sensory and motor nerves, together with their associated tracts and centres in the spinal cord and brain comprise the somatic nervous system, which links the organism with the outside world through perception, awareness and reaction.

Complementary to this is the visceral or autonomic nervous system, which controls the activities of all the internal organs: blood circulation, breathing organs, digestive organs, glands, and so on. The visceral nervous system therefore governs those internal mechanisms that are essential to the life of the body.

The complete nervous system is made up of millions of nerve cells, or neurones, each of which has basically a cell body and a comparatively long nerve fibre. The cell body is a few thousandths of a millimetre thick, whereas the nerve fibre can be more than a metre long. Figure 11 shows a diagrammatic picture of a neurone.

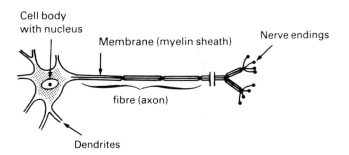

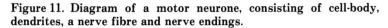

Figure 11. Diagram of a motor neurone, consisting of cell-body, dendrites, a nerve fibre and nerve endings.

15

The function of nerves

The nervous system is essentially a control system, which regulates external and internal movements as well as monitoring a variety of sensations. The working of a neurone depends upon its being sensitive to stimuli, and being able to transmit a stimulus along the length of the nerve fibre. When a nerve cell is stimulated, the resulting impulse travels along the nerve fibre to the operative organ, which may be a muscle fibre, among others.

Nervous impulses are of an electromechanical nature. Nerves are not just 'telephone wires', transmitting impulses passively. A nervous impulse is an active process, self-generating and energy-consuming, more like a length of fuse, or slow match. Unlike a fuse, however, a nerve fibre is not dead once it has been used, but is regenerated in a fraction of a second, becoming receptive once again after the so-called refractory period. The nerve fibre cannot transmit a continuous 'direct current', but only single impulses, with brief interruptions between them.

The speed of transmission is very different in different types of nerve: motor fibres transmit at 70−120 m/s; other fibres in the region of 12−70 m/s.

The nature of nervous impulses

What are these nervous impulses? Like muscle fibres, nerve fibres have a resting membrane potential. *In the resting phase the membrane of the neurone is polarized; positive charges predominate on the outer surface, while negative charges predominate internally. Depolarization of the membrane produces the nervous impulse.* The resting potential, −70 mV, collapses completely, and depolarization continues until a reverse peak of +35 mV is reached. Then comes a repolarization with a quick return to the resting potential of −70 mV. Figure 12 shows the trace of an action potential as a nervous impulse passes along a nerve fibre. As mentioned in Chapter 1, *this electrical oscillation between de- and repolariz-*

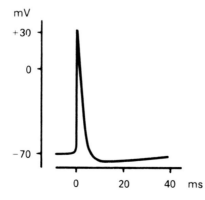

Figure 12. **The action potential in a nerve fibre, which indicates the passage of a nervous impulse.**

ation of the membrane is called the action potential. This is
what nervous impulses are all about.

Causes of the
action potential

The action potential is therefore the electrical manifestation of
a wave of de- and repolarization, which travels along the nerve
fibre at a speed of between 12 and 120 m/s. Depolarization, the
breakdown of the resting potential, is brought about by a
sudden change in the permeability of the membrane, allowing
a stream of positively charged sodium ions to penetrate the
interior of the fibre. Almost simultaneously, positively
charged potassium ions move from inside to outside the
membrane, but these are fewer than the sodium ions moving
inwards, so the net result is a sharp increase in the total
positive charge in the interior. *It is this displacement of*
electrical charges which produces depolarization and the
associated action potential.

In the subsequent repolarization, these charged ions move
in the reverse directions: sodium outwards, potassium
inwards. We are back at the starting point, with the membrane potential restored. This mechanism is called the *sodium-*
potassium pump for short. It is not confined to nerve fibres
but occurs in nearly all living cells. The energy for this
pumping mechanism is derived from adenosine triphosphate
(ATP). The sodium-potassium pump is an essential requirement before a nerve can react to stimuli and transmit
impulses.

The nervous system requires a supply of energy, principally
to maintain the membrane potential, and derives this energy
from ATP. The metabolic activity of a nerve approximately
doubles when it is active; a small increase compared with that
of skeletal muscle, which increases its metabolism 100-fold
when it is working.

Innervation of
muscles

Every muscle is connected to the brain, the overriding control
centre, by nerves of two kinds: efferent or motor nerves, and
afferent, or sensory nerves.

Motor nerves carry impulses, in this case movement orders,
from the brain to the skeletal muscles, where they bring about
a contraction, and collectively control muscular activity.
Within a muscle the nerve divides into its constituent fibres,
each nerve fibre serving to innervate several muscle fibres.
Each single motor neurone, together with the muscle fibres
that it innervates, constitutes a *motor unit*. In muscles which
carry out fine and precise movements, as in skilled work, there
are only 3−6 muscle fibres per motor unit, whereas muscles
which do heavy work may have 100 muscle fibres innervated
by one neurone.

The bundle of motor nerve fibres shown above in Figure 11
ends in so-called *motor end plates*, where the cell membrane
of the nerve fibre is thickened. This is where the motor impulse

leaps across from the nerve fibre to the muscle fibre and where the action potential finally provokes muscular contraction.

Sensory nerves conduct impulses from the muscles into the central nervous system, either to the spinal cord or to the brain. *Sensory impulses are bearers of 'signals'*, which will be utilized in the central nervous system in part to direct muscular work and in part to store as information.

Receptor organs of a special kind are the *muscle spindles*, parallel to the muscle fibres, and ending in both sides of the tendons. The muscle spindles are sensitive to stretching of the muscle, and send signals about this to the spinal cord (proprioceptor system).

Further organs of sensory response are the *organs of Golgi* (Golgi–Mazzoni corpuscles), which are composed of a network of nodular nerve endings and are embedded in the tendons. These organs, too, transmit sensory impulses to the nerve cord whenever the tendon is under tension.

Within the spinal cord the sensory impulses pass by means of an intermediate neurone to the motor nerve, so that new impulses flow back to the muscles. This system of an afferent sensory nerve and an efferent motor nerve running back to the same muscle is called a *reflex arc*. Such a reflex arc keeps muscle tension and muscle length continually adjusted to each other, *the muscle spindle and the organ of Golgi being the 'detectors' in this 'regulatory system'*.

Other sensory nerves conduct impulses from the muscles: over a first intermediate neurone in the spinal cord, and over a second in the medulla of the brain (brain-stem), up to the cerebral cortex *where the incoming pulses are finally experienced as a sensation*, This is how pains which arise in the muscles are felt. The innervation of the skeletal muscles with sensory and motor nerve fibres is demonstrated in Figure 13.

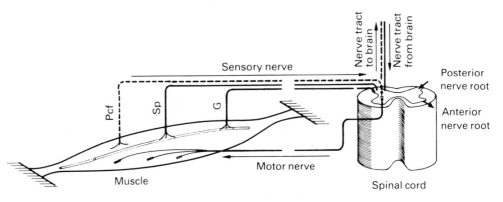

Figure 13. Innervation of a muscle.
The sensory nerve comprises pain-conducting fibres (Pcf) as well as fibres from the muscle spindle (Sp) and from the Golgi receptors of the tendon (G). The fibres of the motor nerve terminate in the motor end plates in the membranes of the muscle fibres.

At the same time the sensory paths leading to the spinal cord and into the brain, as well as the motor paths running back from the brain to the muscles are drawn in very simply.

2.2. Reflexes and skills

One special way in which movement and activity are controlled is by means of reflexes. Because these are not consciously directed they are *'automatic' in a physiological sense*. A reflex consists of three parts: (*a*) an impulse which travels along a sensory nerve, carrying the information to the spinal cord or to the brain; (*b*) intermediate neurones, which pass the impulse across to a motor nerve; and (*c*) a final impulse along the motor nerve which activates the appropriate muscle. An example is reflex blinking of the eyelids. When anything unexpected moves close to either eye, both eyes close automatically. The unexpected movement in the visual field provides the initial stimulus; the sensory impulse travels to a particular centre in the brain, which itself acts as intermediate neurone, and passes on the message to a motor nerve; the motor nerve in turn operates the muscles of the eyelid. Reflex blinking is thus an automatic protective mechanism, safeguarding the eyes against damage. The body makes use of thousands of such reflexes, not only for protection, but as part of normal control functions.

   Reflexes also play an essential part in muscular activity. One example has been described above in connection with skeletal muscle, and the reflex arc involved is shown diagramatically in Figure 13. Another example of a more complicated and important reflex is the antagonistic control of a muscular movement. When the lower arm is bent the bending muscles are caused to contract by stimulation of their motor nerves, and if this is to proceed smoothly the opposing muscles behind the arm must be simultaneously relaxed by exactly the right amount. This is a reflex phenomenon, i.e., automatic direction, by which a trouble-free movement is carried out.

Skilled work

To give an idea of the complexity of nervous control, the most important stages in a piece of skilled work are shown diagrammatically in Figure 14. During a simple grasping operation such as that illustrated, the first step is to make use of visual information to direct the movements of the arms, hands and fingers towards the object to be grasped. For this purpose nervous impulses travel along the optic nerve from the retina of the eye to the brain and are integrated there into a picture; hand—finger—object to be grasped are perceived. These impulses are then passed over to other centres, in the medulla and cerebellum, which control muscular activity. On the

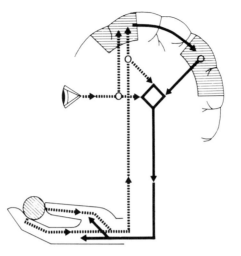

Figure 14. Diagram illustrating nervous controls in skilled work.

strength of the visual signals it has received, the brain thus decides what the next move will be. When the object has been grasped, pressure-sensitive nerves in the skin send new signals to the brain, and the operator can adjust his finger pressure accordingly.

Conditioned reflexes

Becoming skilled is largely a matter of developing new reflexes which go on without conscious control, and are called conditioned reflexes. Figure 14 indicates new reflex pathways as arrows, passing directly from the synapses (intermediate neurones) of the sensory pathways, into the muscular control centres†, where *combinations of movements are retained as a sort of template.* In other words, whenever a sequence of movements is practised for a long time, the *complete movement pattern becomes 'engraved' in the brain.* Co-ordination and delicate adjustment of individual muscular movements are achieved when a continuous stream of sensory information reaches these motor co-ordination centres. Skill reaches a maximum when learning has eliminated conscious control and movements have become automatic. The task of the conscious mind is to concentrate all nervous activity upon the job in hand and to give overriding 'orders' to the control centres.

An example: writing

The whole process of automatism can be illustrated by the example of learning to write. First the child learns to control the movement of hand and fingers so delicately that the correct shapes appear on the paper. Then he begins to make

---

† According to current theory the centres responsible for precise regulation and coordination lie in the basal ganglion of the medulla and in the cerebellum; movement patterns are 'localized' in these areas.

the necessary movements at will. After a very long period of practice the sequence of movements necessary to write each letter has become 'engraved' as a pattern in the motor control centres of his brain. The writing process itself becomes more and more automatic, and eventually the conscious mind concerns itself only with finding the right words and making them into sentences.

3. Improving work efficiency

3.1. Optimal use of muscle strength

General principles

When any form of bodily activity calls for a considerable expenditure of effort the necessary movements must be organised in such a way that the muscles are developing as much power as possible. In this way the muscles will be at their most efficient and most skilful.

Where the work involves holding something statically, a posture must be taken up which will allow as many strong muscles as possible to contribute. This is the quickest way to bring the loading of each muscle down below 15% of its maximum.

Since a muscle is most powerful at the beginning of its contraction it is a good idea, in principle, to start from a posture in which the muscle is fully extended. There are so many exceptions to this general rule, however, that it has more theoretical than practical value. One must also take into account the leverage effect of the bones, and, if several muscles join forces, their total effect. In the last case the force exerted is usually at its greatest when as many muscles as possible contract simultaneously. The maximum force of which a muscle, or group of muscles, is capable depends upon:

Age
Sex
Constitution
State of training
Momentary motivation

Age and sex

Figure 15 sets out the effects of age and sex on muscle strength, according to dates obtained by Hettinger [112]. The peak of muscle strength for both men and women is reached between the ages of 25 and 35 years old. Older workers aged between 50 and 60 can produce only about 75—85% as much muscular strength.

As has already been pointed out in Chapter 1 it can be assumed that the average woman will be only about two-thirds as powerful as the average man.

Hettinger [112] has studied the maximum force developed

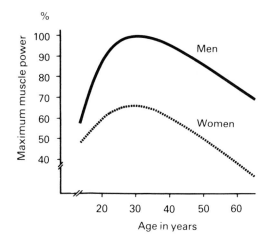

Figure 15. Muscle power in relation to age and sex.
After Hettinger [112].

by both men and women when using three groups of muscles
which he considered have special significance in the assessment of human muscular power. His results are summarized in
Table 2.

Table 2. Maximum muscular force for men and women.
According to Hettinger [112]. SD=*standard deviation of the
individual values.*

Function	Max. force men (N)	SD	Max. force women (N)	SD
Hand clasp	460	120	280	70
Kicking (with knee bent at 90°)	400	60	320	50
Stretching the back	1100	160	740	160

Maximum power when sitting at work

The following rules, deduced from studies made by Caldwell
[38], apply to test subjects who are sitting down, with their
backs against a back rest:

1. The hand is significantly more powerful when it turns
   inwards (pronation) (180 N) than when it turns outwards
   (supination (110 N).
2. This rotation force is greatest if the hand is grasping 30 cm
   in front of the axis of the body.
3. The hand is significantly more powerful when it is pulling
   downwards (370 N) than when pulling upwards (160 N).
4. The hand is more forceful when pushing (600 N) than when
   pulling (360 N).

5. Pushing strength is greatest when the hand is grasping 50 cm in front of the axis of the body.
6. Pulling strength is greatest at a grasping distance of 70 cm.

Maximum force for bending elbows

The maximum force exerted by the muscles which bend the elbow (biceps muscles) seems to be particularly dependent upon leverage in the arm. The results of studies by Clarke [46] and Wakim [265] and their co-workers, as set out in Figure 16, show that *the greatest bending moment is achieved at angles between 90 and 120°.*

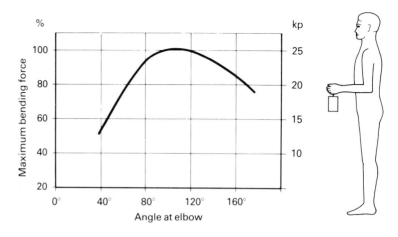

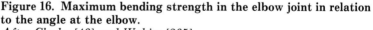

Figure 16. Maximum bending strength in the elbow joint in relation to the angle at the elbow.
After Clarke [46] and Wakim [265].

Maximum force for standing work

The maximum pulling and pushing forces of the hand during standing work have been thoroughly studied by Rohmert [218] and by Rohmert and Jenik [220]. A selection of their results for men are set out in Figure 17.

The following conclusions may be drawn from Rohmert's [218] studies:

1. At most positions of the arms, while standing up, the pushing force is greater than the pulling force.
2. Both pulling and pushing forces are greatest in the vertical plane, and lowest in the horizontal plane.
3. Pulling and pushing forces are of the same order of magnitude whether the arms are held out sideways or forwards (in the sagittal plane)
4. Pushing force in the horizontal plane may amount to:

   160−170 N for men
   80−90 N for women.

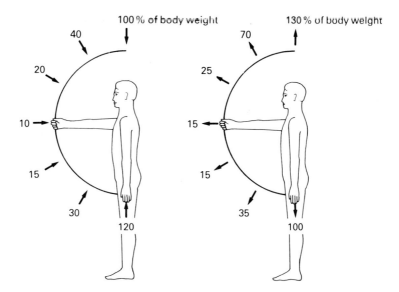

100 % of body weight   130 % of body weight

Figure 17. **Maximum force of pulling (left) and pushing (right) for a man.**
Feet 300 mm apart. The values at various angles are given as a percentage of the body weight. Simplified, after Rohmert [218].

3.2. Practical guidelines for work layout

Most important principles

As has already been emphasized, static loads on the muscles lead to painful fatigue; they are wasteful and exhausting. For this reason, a major objective in the design and layout of jobs, workplaces, machines, instruments and tools should be to *minimize or abolish altogether the need to grasp and hold things.*

Unavoidable static effort should be reduced to not more than 15% of the maximum, and to 8% for long-lasting loads.

According to van Wely [271], dynamic effort of a repetitive nature should not exceed 30% of the maximum, although it may rise to 50% as long as the effort is not prolonged for more than 5 minutes.

Seven guidelines

1. *Avoid any kind of bent or unnatural posture* (see Figure 18). Bending the trunk or the head sideways is more harmful than bending forwards.
2. *Avoid keeping an arm outstretched either forwards or sideways.* Such postures not only lead to rapid fatigue, but also markedly reduce the precision and general level of skill of operations using the hands and arms (see Figure 19).
3. *Work sitting down as much as possible.* Workplaces at which the operator can either stand or sit are to be recommended.

Figure 18. Strongly bent posture while cleaning a casting in a foundry.
If the casting stood up vertically it could be cleaned with substantially less static load on the muscles of the back.

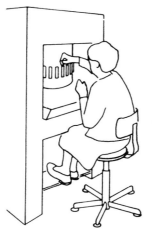

Figure 19. Fitting components onto a machine with the arm continuously outstretched.
The static loads on the right arm and on the shoulder muscles are tiring and reduce skill. The machine should be redesigned so that the operator can work with her elbow lowered and bent at right angles.

4. *Arm movements should be either in opposition to each other or otherwise symmetrical.* Moving one arm by itself sets up static loads on the trunk muscles. Furthermore, symmetrical movements of the arms facilitate nervous control of the operation.
5. *The working field (either the work itself or the table on which it stands) should be at such a height that it is at the best distance from the eyes of the operator.* The shorter his sight, the higher should be the working field (see Figure 20).

Figure 20. The work level should be at such a height that the body takes up a natural posture, slightly inclined forwards, with the eyes at the best viewing distance from the work.
This work bench is a model of its kind, with the elbows supported in a natural position, without static effort.

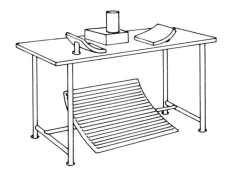

Figure 21. The forearms and elbows can be supported by adjustable, padded rests.
Restful support is also given to the legs by a footrest that is adjustable for height, and which allows all normal leg movements to be made. For women the bars of the footrest should be closer together, to prevent shoe heels becoming trapped.

6. Hand grips, operating levers, tools and materials should be arranged around the workplace in such a way that the most frequent movements are carried out with the elbows bent and near to the body. *The best position for both strength and skill in the hands is to have them 25—30 cm from the eyes, with the elbows lowered and bent at right-angles.*
7. *Hand-work can be raised up by using supports under the elbows, forearms or hands.* These supports should be padded with felt or some other soft, warm material, and should be adjustable to suit people of different sizes (see Figure 21).

4. Problems of body size

Basics

Since natural postures — attitudes of the trunk, arms and legs which do not involve static effort — and natural movements are a necessary part of efficient work, it is also essential that *the workplace should be suited to the body size of the operator.*

Variation

Here we are soon up against a problem: the enormous variations in body size between individuals, the two sexes, and different races. It is not enough, as a rule, to design a workplace to suit an average person. Most often it is necessary to take account of the tallest persons (e.g., to decide leg room under a table), or of the shortest persons (e.g., to make sure they can reach high enough). To take an extreme example: if the heights of doorways were fixed to suit the average person, many people using them would strike their heads on the lintel.

Confidence interval and percentile

Since it is not usually possible to design workplaces to suit the very biggest or the very smallest workers, we must be content with meeting the requirements of the majority. A selection is therefore made of *either 95% or 90% respectively*. A confidence interval (Ci) of 95% means that the smallest $2 \cdot 5\%$ and the largest $2 \cdot 5\%$ are excluded from consideration. Any particular percentage is called a *percentile*: thus in the present example only the percentiles lying between $2 \cdot 5\%$ and $97 \cdot 5\%$ are being considered. If we know the mean ($\bar{x}$) and the standard deviation (SD) of any group measurements then:

Ci 95% = $\bar{x} \pm 1 \cdot 95$ SD
Ci 90% = $\bar{x} \pm 1 \cdot 65$ SD

Figure 22 shows the frequency distribution of a body of American military recruits.

The anthropometric literature of different countries abounds in accounts of human measurements. A most extensive study was recently published by Pheasant [209]. The data in Tables 3 and 4 are taken from this source, where all the references to authors or measurements can be looked up.

Figures 23—28 show the dimensions tabulated by Pheasant; they correspond to the dimensions reported in Tables 3 and 4.

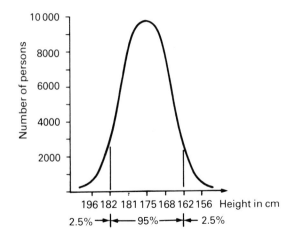
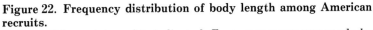

Figure 22. Frequency distribution of body length among American recruits.
95% confidence interval is indicated. From measurements made by Hertzberg et al. [110].

The most important differences in body size are due to sex, age and ethnic diversity.

Since with increasing age people become shorter, but heavier and more corpulent, workplaces should be so designed that they allow for the bodily dimensions of people of all ages between 20 and 65 years. An American survey [261] showed that people in the 45–65-years age group had changed by the

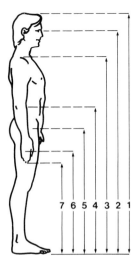

Figure 23. Static anthropometric dimensions (1).
From Pheasant [209].

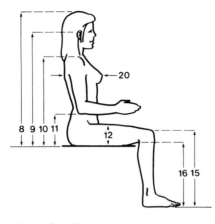

Figure 24. Static anthropometric dimensions (2).
From Pheasant [209].

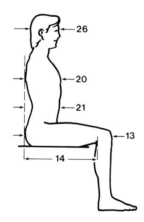

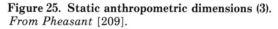

Figure 25. Static anthropometric dimensions (3).
From Pheasant [209].

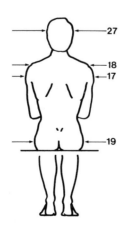

Figure 26. Static anthropometric dimensions (4).
From Pheasant [209]

Figure 27. Static anthropometric dimensions (5).
From Pheasant [209].

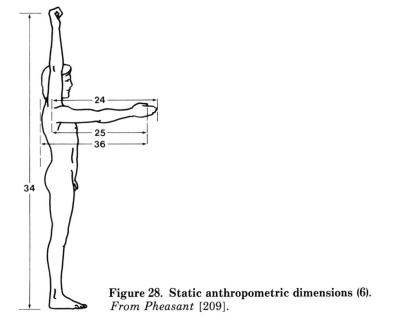

Figure 28. Static anthropometric dimensions (6).
From Pheasant [209].

following amounts compared with the 20-year-age group:

Body length (men and women)   −40 mm
Body weight (men)   +6 kg
Body weight (women)   +10 kg.

Hand size   Hand size is particularly important when designing controls. Figure 29 and Table 5 summarize relevant information according to Jürgens [137].

Table 3. Anthropometric data in mm of British adults aged 19—65 years.
The reference numbers of the dimensions are shown in Figures 23—28.
According to Pheasant [209].

Dimension	Men				Women			
	5th %ile	50th %ile	95th %ile	SD	5th %ile	50th %ile	95th %ile	SD
1. Stature	1625	1740	1855	70	1505	1610	1710	62
2. Eye height	1515	1630	1745	69	1405	1505	1610	61
3. Shoulder height	1315	1425	1535	66	1215	1310	1405	58
4. Elbow height	1005	1090	1180	52	930	1005	1085	46
5. Hip height	840	920	1000	50	740	810	885	43
6. Knuckle height	690	755	825	41	660	720	780	36
7. Fingertip height	590	655	720	38	560	625	685	38
8. Sitting height	850	910	965	36	795	850	910	35
9. Sitting eye height	735	790	845	35	685	740	795	33
10. Sitting shoulder height	540	595	645	32	505	555	610	31
11. Sitting elbow height	195	245	295	31	185	235	280	29
12. Thigh thickness	135	160	185	15	125	155	180	17
13. Buttock-knee length	540	595	645	31	520	570	620	30
14. Buttock-popliteal length	440	495	550	32	435	480	530	30
15. Knee height	490	545	595	32	455	500	540	27
16. Popliteal height	395	440	490	29	355	400	445	27
17. Shoulder breadth (bideltoid)	420	465	510	28	355	395	435	24
18. Shoulder breadth (biacromial)	365	400	430	20	325	355	385	18
19. Hip breadth	310	360	405	29	310	370	435	38
20. Chest (bust) depth	215	250	285	22	210	250	295	27
21. Abdominal depth	220	270	325	32	205	255	305	30
22. Shoulder-elbow length	330	365	395	20	300	330	360	17
23. Elbow-fingertip length	440	475	510	21	400	430	460	19
24. Upper limb length	720	780	840	36	655	705	760	32
25. Shoulder-grip length	610	665	715	32	555	600	650	29
26. Head length	180	195	205	8	165	180	190	7
27. Head breadth	145	155	165	6	135	145	150	6
28. Hand length	175	190	205	10	160	175	190	9
29. Hand breadth	80	85	95	5	70	75	85	4
30. Foot length	240	265	285	14	215	235	255	12
31. Foot breadth	85	95	110	6	80	90	100	6
32. Span	1655	1790	1925	83	1490	1605	1725	71
33. Elbow span	865	945	1020	47	780	850	920	43
34. Vertical grip reach (standing)	1925	2060	2190	80	1790	1905	2020	71
35. Vertical grip reach (sitting)	1145	1245	1340	60	1060	1150	1235	53
36. Forward grip reach	720	780	835	34	650	705	755	31

Table 4. Selected anthropometric data in mm of different nations.
The references number of the dimensions are shown in Figures 23–28. According to Pheasant [209].

Dimension in mm	USA Men 5th %ile	50th %ile	95th %ile	SD	USA Women 5th %ile	50th %ile	95th %ile	SD	France Men 5th %ile	50th %ile	95th %ile	SD	France Women 5th %ile	50th %ile	95th %ile	SD
1. Stature	1640	1755	1870	(71)	1520	1625	1730	(64)	1600	1715	1830	(69)	1500	1600	1700	(61)
3. Shoulder height	1320	1440	1550	(67)	1225	1325	1425	(60)	1300	1405	1510	(65)	1210	1305	1400	(57)
4. Elbow height	1020	1105	1190	(53)	945	1020	1095	(47)	995	1080	1165	(51)	925	1000	1075	(45)
5. Hip height	835	915	995	(50)	760	835	910	(45)	815	895	975	(49)	750	820	890	(43)
8. Sitting height	855	915	975	(36)	800	860	920	(36)	850	910	970	(35)	810	860	910	(31)
9. Sitting eye height	740	800	860	(35)	690	750	810	(35)	735	795	855	(35)	700	750	800	(30)
11. Sitting elbow height	195	245	295	(31)	185	235	285	(29)	190	240	290	(30)	185	230	275	(28)
12. Thigh thickness	135	160	185	(16)	125	155	185	(17)	150	180	210	(17)	135	165	195	(17)
13. Buttock–knee length	550	600	650	(31)	525	575	625	(31)	550	595	640	(28)	520	565	610	(28)
15. Knee height	495	550	605	(32)	460	505	550	(28)	485	530	575	(26)	455	495	535	(24)
16. Popliteal height	395	445	495	(29)	360	405	450	(28)	385	425	465	(25)	350	390	430	(23)
17. Shoulder breadth	425	470	515	(28)	360	400	440	(25)	425	470	515	(26)	380	425	470	(27)
24. Upper limb length	730	790	850	(36)	655	715	775	(35)	710	770	830	(35)	650	705	760	(32)

Dimension in mm	F.R. Germany Men 5th %ile	50th %ile	95th %ile	SD	F.R. Germany Women 5th %ile	50th %ile	95th %ile	SD	Japan Men 5th %ile	50th %ile	95th %ile	SD	Japan Women 5th %ile	50th %ile	95th %ile	SD
1. Stature	1645	1745	1845	(62)	1520	1635	1750	(69)	1560	1655	1750	(58)	1450	1530	1610	(48)
3. Shoulder height	1370	1465	1560	(58)	1240	1320	1400	(50)	1250	1340	1430	(54)	1075	1145	1215	(44)
4. Elbow height	1020	1095	1170	(46)	925	1000	1075	(46)	965	1035	1105	(43)	895	955	1015	(36)
5. Hip height	840	910	980	(44)	760	840	920	(48)	765	830	895	(41)	700	755	810	(33)
8. Sitting height	865	920	975	(32)	800	865	930	(39)	850	900	950	(31)	800	845	890	(28)
9. Sitting eye height	750	800	850	(31)	680	740	800	(37)	735	785	835	(31)	690	735	780	(28)
11. Sitting elbow height	195	235	275	(25)	165	205	245	(23)	220	260	300	(23)	215	250	285	(20)
12. Thigh thickness	135	150	265	(70)	125	155	185	(19)	110	135	160	(14)	105	130	155	(14)
13. Buttock–knee length	560	600	640	(25)	525	580	635	(33)	500	550	600	(29)	485	530	575	(26)
15. Knee height	500	545	590	(28)	455	505	555	(30)	450	490	530	(23)	420	450	480	(18)
16. Popliteal height	415	455	495	(25)	355	395	435	(23)	360	400	440	(24)	325	360	395	(21)
17. Shoulder breadth	425	465	505	(23)	355	400	445	(27)	405	440	475	(22)	365	395	425	(18)
24. Upper limb length	735	785	835	(31)	660	720	780	(36)	665	715	765	(29)	605	645	685	(25)

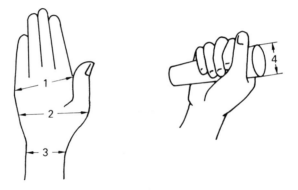

Figure 29. Indication of measurements listed in Table 5.
After Jürgens [137].

Table 5. Dimensions of one hand.
*Mean values (x̄) and 90% confidence interval (Ci) 90% for 8000
20-year-old men, with additional sample checks on women in the
Federal Republic of Germany. After Jürgens [137].*

Measure-ment no.	Bodily dimension (in mm)	Men		Women	
		x̄	Ci 90%	x̄	Ci 90%
1	Circumference of hand	211	193−230	187	175−201
2	Breadth of hand	106	98−113	—	—
3	Circumference of wrist	171	155−188	161	143−179
4	Maximum grip (circumference of thumb and forefinger	134	120−153	—	—

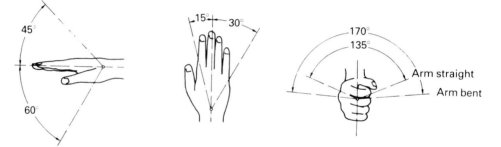

Figure 30. Arcs of movement of the wrist and hand.
After Stier and Meyer [249].

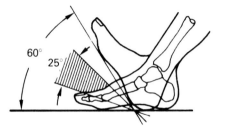

Figure 31. Bending and stretching at the ankle.
Maximum arc 60°; most convenient arc 25—30°. Modified after Stier and Meyer [249].

Angle of rotation of joints

The operating space of a limb is a product of its length and the angle of rotation of its joint. Figure 30 shows some operating spaces available to the hand.

The arm can rotate through an angle of 250° about its axis in the sagittal plane, of which a half-circle (180°) lies in front of the body and a further 70° or thereabouts, backwards [188]. The angle of rotation of the ankle is shown in Figure 31.

5. The design of workstations

5.1. Working heights

Recommenda-
tions are
compromises

The ergonomic recommendations for the dimensions of workstations are only to some extent based on anthropometric data; behavioural patterns of employees and specific requirements of the work itself must also be considered. Thus the recommended dimensions given in textbooks or in various standard works are compromise solutions which may often be quite arbitrary. Another critical remark is necessary: most standard specifications for ergonomic workstations were worked out by committees, in which ergonomics, economics, industry and unions and/or employers were represented. The resulting recommendations seem reasonable and suitable in most cases, but they have seldom been seriously tested under practical conditions. It is therefore not surprising when field studies or practical experience do not always confirm recommended standard dimensions.

Working heights
when standing

Working height is of critical importance in the design of workplaces. If the work is raised too high the shoulders must frequently be lifted up to compensate, which may lead to painful cramps in the neck and shoulders. If the working height is too low the back must be excessively bowed, which again often causes backache. Hence the work table must be of such a height that it suits the height of the operator, whether he stands or sits at his work.

The most favourable working height for handwork while standing is 50—100 mm below elbow level. The average elbow height (distance from floor to underside of elbow when it is bent at right angles with the upper arm vertical) is 1050 mm for men and 980 mm for women.

Hence we may conclude that on average working heights of 950—1000 mm will be convenient for men, and 880—930 mm for women. Besides these anthropometric considerations, we must allow for the nature of the work:

1. For delicate work (e.g., drawing) it is desirable to support the elbow to help to reduce static loads in the muscles of

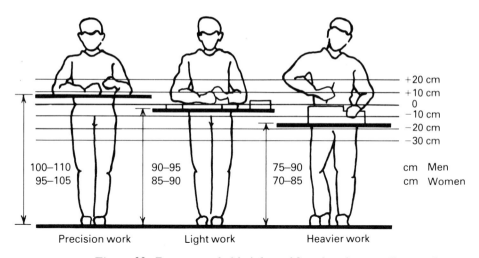

Figure 32. Recommended heights of benches for standing work.
The reference line (± 0) is the height of the elbows above the floor, which averages 1050 mm for men and 980 mm for women.

the back. A good working height is about 50−100 mm above elbow height.

2. During manual work the operator often needs space for tools, materials and containers of various kinds, and a suitable height for these is 100−150 mm below elbow height.

3. During standing work, if this involves much effort and makes use of the weight of the upper part of the body (e.g., woodworking,‘ or heavy assembly work), the working surface needs to be lower; 150−400 mm below elbow height is adequate.

Recommended working heights when standing are set out in Figure 32.

The dimensions recommended in Figure 32 are merely general guidelines, since they are based on average body measurements and make no allowances for individual variation. The table heights quoted are too high for short people, who will need to use wooden foot rests, or some similar support. On the other hand tall people will need to bend over the work table, which will be a cause of backache.

Fully adjustable work tables when standing

Ergonomically speaking, therefore, it is often desirable to be able to adjust the working height to suit the individual. In addition to improvisations such as foot supports, or lengthening the legs of the work table, a fully adjustable bench is to be recommended. Figure 33 shows recommended working heights for light standing work, in relation to the body length of the operators.

If a firm is unable, for financial or practical reasons, to

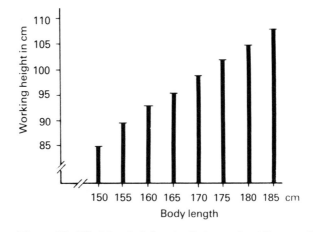

Figure 33. Working heights for light work while standing, in relation to body length.

provide fully adjustable benches, or if the operating level at a machine cannot be varied, then in principle *working heights should be set to suit the tallest operators*: smaller people can be accommodated by giving them something to stand on.

Workplaces for sedentary work

A classic study carried out in 1951 by Ellis [67] is often advanced when working heights are discussed. Ellis was able to confirm an old empirical rule: *the maximum speed of operation for manual jobs carried out in front of the body is achieved by keeping the elbows down at the sides and the arms bent at a right angle.* This is a generally accepted basis for the assessment of working heights for sedentary activities.

Since the work may also require fine or precise manipulation, working heights must take note of the optimum visual distance, as Figure 20 has already shown. In such cases the plane of the work must be raised until the operator can see clearly while keeping his back in a natural posture. The opposite, i.e., lowering of the working surface, is necessary when handwork calls for great force, or much freedom of movement. However, the need for a low table conflicts with the necessity for enough knee room under the table, and this can be a limiting factor. If we extract the measurement 'ground-to-upper surface of knee' for large individuals from Table 3, and if we make certain additions to allow for heels and for a minimum amount of movement, we arrive at *the following figures for free knee room.*

> *men   (590+50 mm)=640 mm*
> *women (540+70 mm)=610 mm*

If we allow 40 mm for the thickness of the table top, assuming it has no rim, *this brings the lowest table level up to*:

> men   680 mm
> women 650 mm

This lowest working surface is recommended for heavy assembly work, for working over any kind of container, and for preparatory work in the kitchen. It should also be noted here that the maximum distance from seating surface to the underside of the table should be 185 mm (95% of the thickness of the thighs).

How office employees sit

A survey carried out in 1962 by Grandjean and Burandt [86] on 261 men and 117 women engaged in traditional office work revealed interesting links between desk height and musculo-skeletal troubles. The work-sampling analysis gave particulars about the different sitting postures shown in Figure 34.

	Sitting forward on chair	15%
	Sitting in middle of chair	52%
	Sitting back on chair	33%
	Leaning on backrest	42%
	Arms resting on table	40%

Figure 34. Sitting postures of 378 office employees, as shown by a multimoment observation technique.
4920 observations. The percentages quoted indicate how much of the working period was spent in that posture. The two lower observations were seen simultaneously with the three upper postures; that is why the sum of all 5 characteristics exceeds 100%. After Grandjean and Burandt [86]

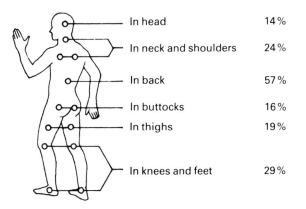

In head	14%
In neck and shoulders	24%
In back	57%
In buttocks	16%
In thighs	19%
In knees and feet	29%

Figure 35. Incidence of bodily aches among 246 employees engaged in traditional sedentary office jobs.
Multiple answers were possible. After Grandjean and Burandt [86].

An upright trunk posture was observed for only about 50% of the time, the trunk leaning against the backrest about 40% of the time, although most of the chairs were provided with rather poor back supports.

Figure 35 presents the results of the survey on musculo-skeletal complaints.

Links between desk height, sitting behaviour and pains

The principal anthropometric data and desk heights were assessed and compared with the reports on musculoskeletal troubles. From a large number of results and calculated correlations, the following conclusions emerged:

1. 24% reported pains in neck and shoulders which most of the subjects, especially the typists, blamed on a too high desk top.
2. 29% reported pains in the knees and feet, most of them short people who had to sit on the front edge of their chair, probably because they had no footrests.
3. *A desk top 740—780 mm high gave the employees the most scope for adaptation to suit themselves,* provided that a fully adjustable seat and foot rests were available.
4. *Regardless of their stature, the great majority of workers preferred the seat to be 270—300 mm below the desk top. This seems to permit a natural position of the trunk, obviously a point of major importance to these employees.*
5. The incidence of backache (57%) and the frequent use of the backrest (42% of the time) indicate the need to relax the back muscles periodically and may be quoted as evidence of the importance of a well-constructed backrest.

Electromyography of shoulder muscles

Several authors have recorded the electrical activity of the shoulder muscles while subjects worked at different desk heights. It should be remembered here that the electrical activity of a muscle is an indicator of the exerted muscular force; the procedure is called electromyography. As far back as 1951, Lundervold [171] investigated the electrical activity of shoulder and arm muscles of subjects operating typewriters at high and lower levels. (At that time electrical impulses were obtained with ink-writer records!). The author concluded: '... the smallest number of action potentials (electrical impulses) were recorded when the person undergoing the experiment was sitting in a relaxed and well balanced state of equilibrium, or was using a backrest'. High-level typewriting was associated with raised shoulders and a strongly increased electrical activity of the trapezius and deltoid muscles. (The trapezius muscle lifts the shoulders, the deltoid the upper arms). Recently, Hagberg [99] made a quantitative analysis of the electromyograms of shoulder and arm muscles when typing at different heights. These results are shown in Figure 36.

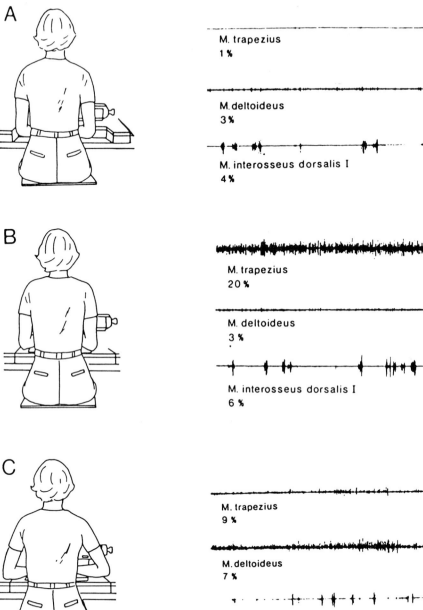

Figure 36. Electromyographic recording of shoulder muscle activity. *The figures in brackets refer to percentage of maximum voluntary contraction position. A = optimal height of the typewriter, (i.e., home row at elbow height). B = too high, resulting in elevation of shoulders by the trapezius muscle. C = too high, compensated by a sideward elevation of the upper arms by the deltoid muscle. According to Hagberg [99].*

Lifting the shoulders is strenuous static work

Thus a too high working level can be compensated either by lifting the shoulders through contraction of the trapezius muscle or by lifting the upper arms with the deltoid muscle. Moreover, the contraction force of the shoulder-lifting muscle reaches 20% of the maximum force, which would certainly suffice to eventually generate great pains in the shoulder muscles.

Most important: seat to desk distance

A slight forward stoop, with the arms on the desk, is only minimally tiring when reading or writing, but in order to relax the back *the distance from seat surface to desk top must be between 270 and 300 mm.* As mentioned above, employees sitting at an office desk first of all look for a comfortable and relaxed trunk position, and often accept a seat height that is bad for the legs or buttocks rather than sacrifice a comfortable trunk posture.

The height of tables which are not adjustable is primarily based on average body measurements and makes no allowance for individual variation. Therefore all table heights recommended are too high for short people, who will need some kind of footrest. On the other hand, tall people will have to bend the neck over the work table which will cause musculoskeletal troubles in the neck and back. Table heights for traditional office work (with the exception of typing!) thus follow the rule that it is more practical to choose a height to suit the tall rather than the short person; the latter can always be given a footrest to raise the seat to a suitable level. On the other hand, a tall person given a table that is too low can do nothing about it except fix the seat so low that it might be uncomfortable for the legs.

Conclusion for an office desk without typewriter

As illustrated in Figure 37, *office desks to be used without a typewriter should have a height of 740—780 mm for men and 700—740 mm for women, assuming that the chairs are fully*

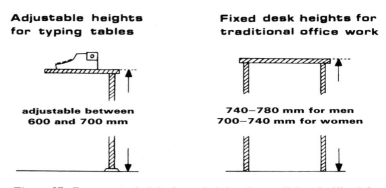

Adjustable heights for typing tables

adjustable between 600 and 700 mm

Fixed desk heights for traditional office work

740—780 mm for men
700—740 mm for women

Figure 37. Recommended desk-top heights for traditional office jobs.
Left: Range of adjustability for typing desks.
Right: Desk-top heights for reading and writing without typewriter.

adjustable and footrests are available. These figures are slightly higher than most standard specifications recommending desk top heights between 720 and 750 mm, which are certainly not ideal for tall male employees.

Vertical and horizontal leg room

It is important that office desks allow plenty of space for leg movement and it is an advantage if the legs can be crossed without difficulty. For this reason there should be no drawers above the knees and no thick edge to the desk top. The table top should not be thicker than 30 mm *and the space for legs and feet under the table should be at least 680 mm wide and 690 mm high.*

Many employees, especially those who lean back, periodically like to stretch their legs under the table. It is therefore necessary to leave enough space in the horizontal plane as well. *At knee level the distance from table edge to back wall should not be less than 600 mm and at the level of the feet at least 800 mm.* These recommendations are also valid for workplaces with typewriters or VDTs.

Typing desks call for height adjustability

The above mentioned classical guideline, requiring a straight upright trunk position with the elbows at the sides and bent at right angles has for a long time been the basis for the design of typing desks. Since the height of the keyboard defines the working level, the middle row (or so-called home row) should be at about elbow level. However, the need for such a low desk conflicts with the necessity for enough knee space under the table. This can be a limiting factor and *calls for height-adjustable typing desks*. Indeed, *today most experts recommend the height of such desks to be adjustable between 600 and 700 mm*. These recommendations are illustrated in Figure 37.

Recommendations for fixed typing desks are problematic

The height of a non-adjustable typing desk is a very problematic dimension. *Until now most experts have recommended a fixed top height of 650 mm.* This recommendation is based on two assumptions which are questionable for the following reasons:

1. The assumed straight upright trunk posture is normally not adopted by typists if the work lasts for a few hours and more. They often lean back or forwards in order to relax the back muscles or to get a suitable viewing distance.
2. Old mechanical typewriters required stroke forces of more than 5 N. Modern electrical typewriters need much lower stroke forces of 0·4 to 0·8 N; these keys are easily operated by the fingers alone and no dynamic muscular effort is required from the forearms and hands, which therefore need not be kept in a horizontal plane!

Fitting the task to the Man

Table 6. Desk heights (in mm) for seated work.

Kind of work	Men	Women
Precision work with close visual range	900−1100	800−1000
Reading and writing	740−780	700−740
Manual work requiring strength or space for containers	680	650
Adjustable range for typing desks	600−700	600−700

Thus it is concluded that the recommended height of 650 mm for a fixed typing desk has yet to be demonstrated in a convincing manner by taking into account the preferred and actually assumed postures of typists and by providing operators with comfortable office chairs as well as with electrical typewriters.

Concluding, one can at present recommend the desk heights reported in Table 6.

Alternate sitting and standing

A workplace which allows the operator to sit or stand, as he wishes, is to be highly recommended from a physiological and orthopaedic point of view. There is less need to hold things when one is sitting down than when standing up; on the other hand sitting is the cause of many aches and pains which can be relieved by standing. Standing and sitting impose stresses upon different muscles, so that each changeover relaxes some muscles and stresses others. Furthermore we have good grounds for believing that each change from standing to sitting (and vice versa) is accompanied by variations in the supply of nutrients to the intervertebral discs [146], so that the change is beneficial in this respect, too.

Figure 38 shows a machine which permits alternate sitting and standing, and the relevant measurements are as follows:

Horizontal knee room	300 × 650 mm
Height of working field above seat	300− 600 mm
Height of working field above floor	1000−1200 mm
Range of adjustment of seat	800−1000 mm

Inclined desk top and trunk posture

The question of work surfaces for special purposes arises in relation to school desks and to tables where much reading and writing is carried on. Eastman and Kamon [65] made a contribution to this problem. Six male subjects carried out writing and reading exercises for $2\frac{1}{2}$ hours each at a flat table, and at ones which were tilted to 12° and 24° respectively. Inclinations of the body (measured as the angle between the horizontal plane and a straight line from the twelfth thoracic vertebra to the eyes), were as follows:

Flat table	35−45°
Table tilted 12°	37−48°
Table tilted 24°	40−50°

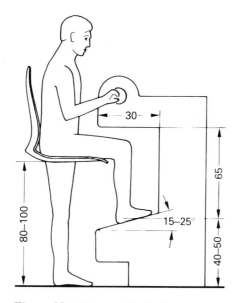

Figure 38. **Alternately sitting and standing at work.**

So, tilting the table top leads to a more erect posture, as well as to fewer pulses on the electromyogram. Subjectively, there was an impression of fewer aches and pains.

In a recent study Bendix and Hagberg [14] examined the effects of inclined desk tops on 10 reading subjects. With increasing desk inclination the cervical as well as the lumbar spine were extended and the head and trunk assumed a more upright posture. The electromyography of the trapezius muscle (shoulder-lifting muscle) showed no change in muscle strain. A rating of acceptability for both reading and writing with either desk inclination favoured a steep slope of the desk for reading but a flatter slope for writing. The authors conclude that reading material should be placed on a sloping desk and paper for writing on a horizontal tabletop. A separate sloping desk placed on a level table should be preferred to inclining the total tabletop since a slope of more than 10° usually causes paper and pencils to slide off.

As a general rule it seems that a tilted work table is an improvement, posturally as well as visually. These advantages have certain drawbacks, however, notably that it is difficult to put things down on an inclined surface.

5.2. Neck and head postures

The posture of the neck and head is not easy to assess since seven joints determine the mobility of this part of the body. In fact, it is possible to combine an erect or even backward flexed (lordotic) neck with a downward bent head, or a forward flexed neck with an upwards directed head. Some authors define the neck/head posture by measuring an angle between a line along the neck related either to a horizontal (or vertical) or to a line along the trunk. Most authors consider such a neck/head angle of 15° to be acceptable. Chaffin [42] determined the neck/head angle related to the horizontal and observed on five subjects that the average time to reach a marked muscle fatigue was shorter with increasing neck/head angles. The author concludes that localized muscle fatigue in the neck area can be a preliminary sign of other more serious and chronic musculoskeletal disorders and that the inclination angle of the head should not exceed 20−30° for any prolonged period of time. Let us note here that neck pains can vanish overnight when source documents are presented on raised reading stands instead of laying them flat on the desktop.

Another way of assessing neck/head posture is by measuring the angle between a line running from the seventh cervical vertebra to the earhole and a vertical line.

'Normal' line of sight ...

A further approach to the problem of neck/head postures is the assessment of the 'normal' line of sight. The direction of sight is determined first by the movement of the eyeballs and secondly by the posture of the neck and head. Eye movement within 15° above and below the normal line of sight is still comfortable. That means that regular viewing tasks should be within a 30° cone around the principal line of sight. If a target lies outside this cone it is assumed that the neck/head mechanism is involved.

... is 10−15° below a horizontal plane

What is the 'normal' line of sight? Most authors today agree that the 'normal' line of sight is some 10−15° below the horizontal plane. Although not stated explicitly, most authors consider the 'normal' line of sight a resting condition of the eye. *For this reason a display or any other visual target should be placed within a viewing angle between 5° above and 30° below a horizontal plane.* This may refer to the eye rotation as such or merely to the slight inclination of the head. The 'normal' line of sight with the cone of easy eye rotation is shown in Figure 39.

Two recent studies confirmed that the 'normal' line of sight is also the preferred line of sight of operators watching a screen. VDT operators who could fix the height of the screen as desired came up with an average figure for the preferred viewing angle of 9° below the horizontal [94]. These results are reported in Table 10 and will be discussed in detail in Chapter 5.

In a recent study, carried out by Bhatnager, Drury and Schiro [16], performance, discomfort and posture were analysed in relation to different height positions of a screen. Twelve

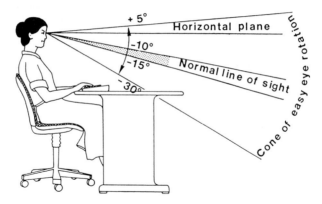

Figure 39. The normal line of sight with the range of easy eye rotation.

subjects inspected printed circuit boards for 3 hours and had to detect breaks in the circuit. They had sight line angles of 23° down, 10° up and 38° up, which determined postures and screen heights. The best performance measure was for the screen height closest to a downward sight line of 15°.

As working time went on, subjects began to lean forward, changed position more often, remarked more about discomfort, and the quality of their performance deteriorated. The screen height had an effect both on frequency of posture changes (a sensitive indicator of postural fatigue) and on degree of discomfort. The least comfortable position and the worst performance were effected by the high screen position. This was farthest away from a 10° downward line of sight. Performance and physical comfort were best for the middle screen height closest to the 'normal' line of sight.

Differing opinions

A 'normal' line of sight of 10−15° below the horizontal contradicts an earlier recommendation of Lehmann and Stier [170], who found that seated subjects preferred an average viewing angle of 38° below the horizontal, whereby approximately half of this angle was attributed to the downward bent head. This controversial result is most probably due to special experimental conditions including short test durations.

Line of sight related to head position

In a recent study Hill and Kroemer [118] investigated the preferred line of sight by measuring its angle to the so-called Frankfurt plane, which defines the spatial orientation of the head by anthropometric landmarks. The Frankfurt plane is defined by a line passing through the right earhole and the lowest edge of the right orbit. Naturally this plane moves only with the head. It coincides with the horizontal when the head is held straight and erect. Thirty-two subjects were tested in several experiments, with a total duration of about 2 hours. The seat was provided with a high backrest and could be inclined backwards from 90° to 105° and 130°. The subject's head was placed against the headrest and was not flexed. The results showed an overall mean angle of 34° below the Frankfurt plane with a large range between 14° above and 71° below. When the subjects leaned back the Frankfurt plane was raised by 15° and 40°, respectively.

Relating the line of sight to a horizontal (and not to the Frankfurt plane) yields the following results:

	Line of sight
Backrest angle	related to horizontal
90° (upright posture)	28·6° below
105° (leaning back)	19·4° below
130° (leaning back)	8·0° above

These results are a little confusing; they are not in agreement with the general opinion of a 10° or 15° declination of the line of sight against the horizontal. Again it must be assumed that the special experimental conditions and the relatively short duration of each test might to some extent be responsible for these deviating results.

Conclusions for posture of neck and head

The present state of knowledge suggests that head and neck should not be bent forward by more than 15°, otherwise fatigue and troubles are likely to occur. The preferred line of sight lies on average between 10 and 15° below a horizontal plane and this corresponds very well to the preferred viewing angles of VDT operators watching their screen.

5.3. Room to grasp and move things

Vertical grasp

An understanding of how much room the hands and arms need to take hold of things and to move them about is an important factor in the planning of controls, tools, accessories of various kinds and places on which to put these down. Reaching too far to pick things up leads to excessive movements of the trunk, making the operation itself less accurate, and increasing the risks of pains in the back and shoulders.

Vertical grasp means the radius of action of the arms, with the hands in a grasping attitude. A decisive factor is the height of the axis of the shoulder joint, and the distance from this joint to the closed hand. This is a case in which we need to consider the 5th percentiles of short people, whose measurements are as follows:

Shoulder height when standing	
Small men	1315 mm
Small women	1215 mm
Shoulder height above seat	
Small men	540 mm
Small women	505 mm

Arm length (closed hand)
Small men   610 mm
Small women   555 mm

The vertical grasp in the sagittal plane of the body is in practice an arc of radius 610 mm (men) and 550 mm (women) above shoulder height. If we also allow for lateral movements of the arm the grasping space becomes a semi-circular shell of comparable radius. Figure 40 shows vertical grasp in the sagittal plane.

An occasional stretch to reach beyond this range is permissible since the momentary effect on trunk and shoulders is transient.

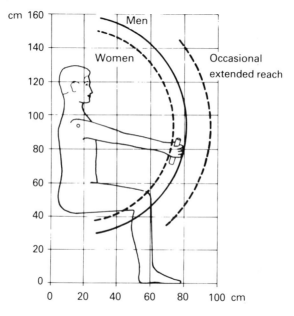

Figure 40. Arc of vertical grasp in the sagittal plane.
The figures include the 5th %ile and so apply to small men and women. The grasp can be extended occasionally by reaching out another 150 mm or so, by stretching the feet, legs and shoulders.

Height of reach

For the positioning of shelves, storage surfaces, special hand grips and other controls it is necessary to know what is the maximum possible reach upwards. Thiberg [254] has calculated regression lines and correlation coefficients for the ratio between body length and height of reach in both men and women. His results when a hand is extended towards a shelf are illustrated in Figure 41.

The following is the ratio between height of reach and body length:

maximum height of reach $= 1·24 \times$ *body length*

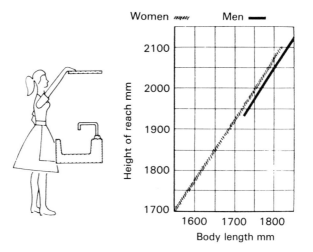

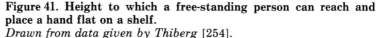

Figure 41. Height to which a free-standing person can reach and place a hand flat on a shelf.
Drawn from data given by Thiberg [254].

Table 7 summarizes maximum reach and vertical grasp for men and women.

It is often necessary to be able to look to the back of the shelf, in order to see the object to be grasped: a bottle, a box or some other container. Under these conditions the highest shelf should be not more than:

1500−1600 mm for men
1400−1500 mm for women

At these heights we can allow a further 600 mm for the depth of the shelf.

Table 7. Maximum height of reach for men and women.

	Percentile	To fingertip (mm)	Grasping height (mm)
Men			
Tall	95	2310	2190
Average	50	2180	2060
Short	5	2040	1920
Women			
Tall	95	2120	2020
Average	50	2000	1900
Short	5	1890	1790

Horizontal grasp at table top level   Figure 42 illustrates grasping and working space over a table top. All materials, tools, controls and containers should be deployed within this space, remembering that an occasional

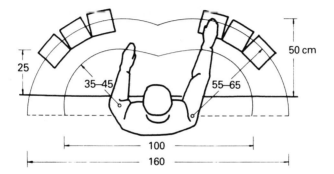

Figure 42. Horizontal arc of grasp, and working area at table top height.
The grasping distance takes account of the distance from shoulder to hand; the working distance only elbow to hand. The values include the 5th %ile, and so apply to men and women of less than average size.

stretch up to distances of 700–800 mm can be undertaken without ill-effects.

Operating space for the legs

Room to operate the feet is essential for all forms of pedal control. This is illustrated in Figure 43.

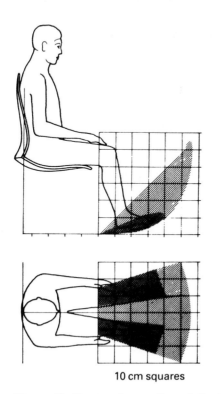

10 cm squares

Figure 43. Range of operation of the feet.
The optimal range for delicately controlled pedal work, where only slight force is needed, is double-shaded. After Kroemer [149].

Kroemer [149] carried out a very thorough investigation into the optimal placing and construction of pedals, to which we shall return in Chapter 9.

5.4. Seating at work

Historical background

Some primitive peoples have no knowledge of seats of any kinds; they crouch, kneel or squat. So where did seats come from? The anthropological answer is that they originated as status symbols; only the chief had the right to be raised up.

Figure 44. The seat of the mayor of Berne (Switzerland) created by cabinet-maker M. Funk in 1735.
This beautiful chair was certainly, above all, a status symbol for the distinguished mayor who was allowed to use it. (Historisches Museum Berne, Switzerland.)

Hence the gradual development of ceremonial stools, which indicated status by their size and decoration. The highest point of this development in cultural history is the splendour of a throne. Figure 44 shows such an example.

This status function persists to the present day. Anyone who doubts this should look inside a factory making office furniture, and notice that there is a type of chair appropriate to each salary level. Thus there is, for example, a wooden chair for typists; for the senior clerk one that is thinly upholstered; a thickly upholstered chair for the office manager, and for directors a swivel armchair upholstered in leather.

At the beginning of the present century the idea gradually emerged that well-being and efficiency are improved and fatigue reduced if people can sit at their work. The reason is a physiological one. As long as a person is standing it requires an outlay of static muscular effort to keep the joints of the feet, knees and hips in fixed positions; this muscular effort ceases when the person sits down.

This realization led to a greater application of medical and ergonomic ideas to the design of seats for work. This development gained in importance as more and more people sat down at their work, until today about three-quarters of all operatives in industrial countries have sedentary jobs.

Pros and cons

The advantages of sedentary work are:

1. Taking the weight off the legs.
2. Ability to avoid unnatural body postures.
3. Reduced energy consumption.
4. Fewer demands on the blood system.

These advantages must be set against certain drawbacks. Prolonged sitting leads to a slackening of the abdominal muscles ('sedentary tummy') and to curvature of the spine, which in turn is bad for the organs of digestion and breathing.

Main problem is the back

The most severe problem involves the spine and the muscles of the back, which in many sitting positions are not merely not relaxed, but are positively stressed in various ways.

About 60% of adults have backache at least once in their lives, and the commonest cause of this is disc trouble.

Intervertebral discs

An intervertebral disc is a sort of cushion, which separates two vertebrae and collectively they give flexibility to the spine. A disc consists internally of a viscous fluid, enclosed in a tough, fibrous ring, which encircles the disc. A schematic representation of a disc between two vertebrae and its connections with the spinal cord and the nervous tracts is given in Figure 45.

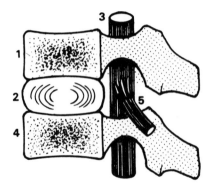

Figure 45. Diagram of a section of the spine.
*The disc (2) lies between two vertebrae, (1) and (4); behind, the spinal
cord (3) and a nervous tract (5). The disc is like a cushion which gives
flexibility to the spine.*

Disc injuries

For reasons that are still unknown, intervertebral discs may
degenerate and lose their strength: they become flattened and
in advanced cases the viscous fluid may even be squeezed out.
The degenerative processes impair the mechanics of the vertebral column and allow tissues and nerves to be strained and
pinched, leading to various back troubles, most commonly
lumbago (painful muscle cramps), and sciatic troubles, and
even in severe instances to paralysis of the legs.

Unnatural postures, weight-lifting and bad seating can
speed up the deterioration of the discs, resulting in all the
ailments mentioned above. For this reason many orthopaedists have started to concern themselves with the medical
aspects of the sitting posture; they include Akerblom [3],
Schoberth [235], Yamaguchi [282], Keegan [141], Nachemson
[196], Andersson and Ortengren [5] and Krämer [146].

*Orthopaedic
research*

A very important contribution was made by the Swedish
orthopaedists Nachemson [196] and Andersson [5], who
developed sophisticated methods to measure the pressure
inside a disc during a variety of standing and sitting postures.
They emphasize that increased disc pressure means that the
discs are being overloaded and will wear out more quickly.
Therefore disc pressure is a criterion for evaluating the risk of
disc injuries and backaches.

*Disc pressure for
four postures*

The effects of four different postures on nine healthy subjects
are shown in Figure 46. The results disclose *that the disc
pressure is greater when sitting than when standing.* The
explanation lies in the mechanism of the pelvis and sacrum
during the transition from standing to sitting:

> The upper edge of the pelvis is rotated backwards
> The sacrum turns upright
> The vertebral column changes from a lordosis to either a
> straight or a kyphotic shape.

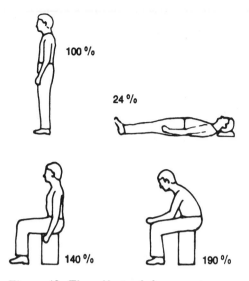

Figure 46. The effect of four postures on the intervertebral disc pressure between the 3rd and 4th lumbar vertebrae.
The pressure measured when standing is taken as 100%. According to Nachemson and Elfström [196].

The spine when standing and sitting

It should be recalled here that lordosis means that the spine is curved forwards, as it normally is in the lumbar region when standing erect. Kyphosis describes the backward curving, which is normal in the thoracic region when standing upright.

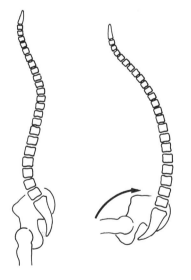

Figure 47. Rotation of the pelvis when changing from a standing to a sitting posture.
Left: Standing upright
Right: Sitting down. Sitting down involves a backward rotation of the pelvis (indicated by the arrow), bringing the sacrum to an upright position and turning the lumbar lordosis into a kyphosis.

These effects of the standing and sitting posture are illustrated in Figure 47.

The backward rotation of the pelvis puts the spine into a state of kyphosis at this point which, in turn, increases the pressure within the discs.

Erect or relaxed sitting posture

That is the reason why many orthopaedists still recommend an upright sitting posture, as this holds the spine in a shape of an elongated S with a lordosis of the lumbar spine. In fact, disc pressure is lower in such a posture than when the body is curved forwards with a predominant kyphosis in the lumbar and thoracic part. One of the recent advocates for an upright trunk at working desks is Mandal [177] who recommends higher seats and higher sloping desks which automatically leads to a more upright posture with reduced forward bending of the back.

In the same line are the *Balans* seats from Norway, which induce a half-sitting—half-kneeling posture. The seat surface is tilted forward (24°), and a support for the knees prevents a forward sliding of the buttocks. The result is a wide opening of the hip angle (between legs and trunk) and a pronounced lumbar lordosis with a straight upright trunk posture. Krueger [152] tested four models and found that the load on knees and lower legs is too high and sitting becomes painful after a while. (Some subjects even refused to sit longer than 2 hours). With desks of 720—780 mm height the effect of a lumbar lordosis drops since the subjects cannot avoid bending the trunk forward. Drury and Francher [59] tested a similar forward-tilting chair. It elicited mixed responses, with complaints of leg discomfort from VDT users. Overall, the chair was no better than conventional chairs and could be worse than well-designed office seats. Looking at these forward-tilting chairs, the question arises whether they would not be of greater use in a physiotherapeutical exercise than in an office!

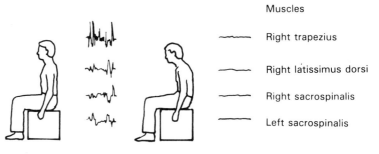

Muscles

——— Right trapezius

——— Right latissimus dorsi

——— Right sacrospinalis

——— Left sacrospinalis

Figure 48. Electrical activity in the back muscles when sitting upright, and in a relaxed posture, slightly bent forward.
Sitting upright involves a considerable electrical activity, revealing the static effort imposed upon the muscles of the back. According to Lundervold [171].

The orthopaedic advice of an upright trunk posture conflicts with the fact that a slightly forward or reclined sitting posture relieves the strain on the back muscles and makes sitting more comfortable. This was established, in part, through electromyographic studies made by Lundervold [171]. Some of his results are shown in Figure 48.

Slightly forward bent trunk holds body weight in balance

From this it emerges that a relaxed posture, with a slightly forward bent trunk holds the weight of the body in balance with itself. This is the posture that many people adopt when they make notes or read in a sitting position, because it is relaxing and exerts a minimum of strain on the muscles of the back. The visual distance for good readability might in some cases be another reason for a slightly forward bent trunk. Thus there is a 'conflict of interests' between the demands of the muscles and those of the intervertebral discs. While the discs prefer an erect posture, the muscles prefer a slight forward bending.

'Feeding the intervertebral discs'

Here we must again refer to the interesting work of Krämer [146], who has made a close study of the nutritional needs of intervertebral discs. The interior of a disc has no blood supply, and must be fed by diffusion through the fibrous outer ring. Krämer has produced evidence that pressure on the disc creates a diffusion gradient from the interior to the exterior, so that tissue fluid leaks out. When the pressure is reduced, this gradient is reversed, and tissue fluid diffuses back, taking nutrients with it. It seems from this that, to keep the discs well nourished and in good condition, they need to be subjected to frequent changes of pressure, as a kind of pump mechanism.

From a medical point of view, therefore, *an occasional change of posture from bent to erect, and vice versa must be beneficial.*

Increased seat angle reduces disc load

The above-mentioned orthopaedists Andersson and Ortengren [5] studied the effects of seat angle and different postures at desks on disc pressure. The electrical activity of back muscles was recorded to measure the static load. The effects of different postures are described in Figure 49. The results show that both leaning back and bending forward with supported upper limbs (writing posture) are favourable conditions for the disc pressure.

The effects of the seat angle are represented in Figure 50. The results are clear: by increasing the seat angle both disc pressure and muscle strain are reduced.

A proper lumbar pad relieves disc strain

Another study by the same authors [5] showed that a proper lumbar support also resulted in a decrease in disc pressure. These results are reported in Figure 51.

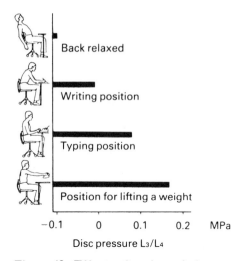

Figure 49. Effects of various sitting postures on disc pressure.

Figure 49. Effects of various sitting postures on disc pressure.
(1 MPa=10·2 kp/cm².) L_3 and L_4=third and fourth lumbar ver-
tebrae. Zero (0) on the pressure scale is a relative reference value
for a seat angle of 90°. Absolute values at the reference level zero
were about 0·5 MPa (= 5 kp/cm²). According to Andersson and
Ortengren [5].

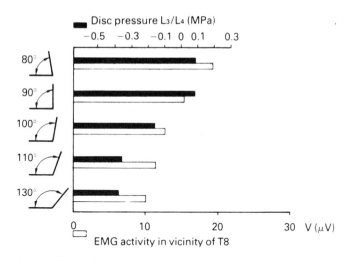

Figure 50. Effect of seat angle (i.e., between seat and backrest) on
disc pressure and on the electrical activity in the back muscles
recorded at the level of the eighth thoracic vertebra (T_8).
For more details see Figure 49. According to Andersson and Orten-
gren [5].

Further studies on the adjustment of back support of office
chairs at different lumbar levels showed that fixing the back
support at the level of the fourth and fifth lumbar vertebrae
slightly decreased pressure compared to placing it at the first
and second. The use of arm rests always resulted in a decrease
in disc pressure, less pronounced, however, when the back-

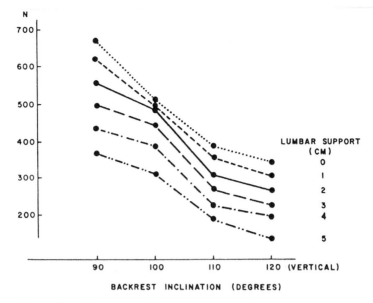

Figure 51. Effects of different sizes of lumbar support and of increasing seat angles on disc pressure.
The size of the lumbar support is defined as the distance between the front of the lumbar pad and the plane of the backrest. The backrest inclination is defined as the angle between the seat and the backrest. From Andersson et al. [5].

rest—seat angle was large. *A comparison between these findings shows that the disc load of a person leaning back with an angle between 110° and 120° and supplied with a 50 mm lumbar pad is even lower than that of a standing posture with the advocated lordosis of the lumbar region.*

Conclusion from orthopaedic research

All these studies lead to an important conclusion: *resting the back against an inclined backrest transfers a relevant portion of the weight of the upper part of the body to the backrest and reduces the strain on discs and muscles. In view of the design of chairs it is deduced that optimum conditions concerning disc pressure and muscular activity are given when the backrest has an inclination of 110° or 120° and a lumbar pad 50 mm thick.*

The cervical spine

Another part of the spine is as important as the lumbar region: the cervical spine or spine of the neck, consisting of the first seven vertebrae. Like the lumbar spine, this is a very mobile segment, also showing a lordosis when standing upright. The cervical spine is a delicate part and prone to degenerative processes and arthrosis. A great number of adults have neck troubles due to injuries of the cervical vertebrae and discs, generally referred to as cervical syndrome. The most common symptoms of cervical syndrome are painful cramps in the shoulder muscles, pains and reduced mobility in the cervical spine and sometimes painful radiation in the arms, ailments which are also called cervicobrachial syndrome. In Japan these cervicobrachial disorders are considered an occupational disease since they are often observed among key-punchers, assembly-plant workers, typists, cash register operators and telephone operators [175]. Recently, several authors have discovered physical troubles among VDT operators which fit the above description of cervicobrachial disorders [156, 203]. It is often observed that discomfort in the neck increases with the degree of forward bending of the head.

Ergonomic research

A 'sitting-machine' [90] and a variety of moulded seat shells [92] of different profiles were tested on a large number of people, including a group of 68 who complained of back ailments. They were asked to give their subjective impressions of the different seat profiles, and their effects upon various parts of the body. The profiles of a multipurpose seat and an easy-chair, which the test subjects voted to produce the fewest aches and pains, are shown in Figure 52.

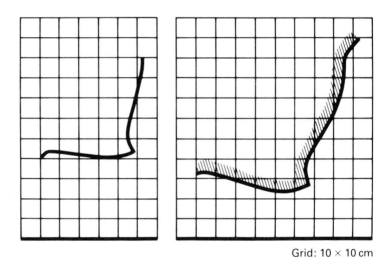

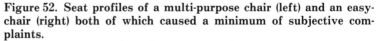

Grid: 10 × 10 cm

Figure 52. Seat profiles of a multi-purpose chair (left) and an easy-chair (right) both of which caused a minimum of subjective complaints.
Grid 100×100 mm. After Grandjean et al. [90, 92].

The best chair for relaxing

The result for the easy-chair agrees almost exactly with what the orthopaedists say. *A seat profile which produces a low pressure in the intervertebral discs and requires very little static muscular effort is also the one that causes the fewest aches and pains. When more discomfort is experienced, it is evidently associated with stresses falling upon the discs and fatigue symptoms in the muscles.* This link is also pointed out by Rosemeyer [222], who noted that opening out the angle between the seat and backrest to 110° resulted in less electrical activity in the muscles and greater comfort.

Taking these researches as a whole, the following recommendations for an easy-chair have both orthopaedic/medical and ergonomic backing:

1. *The seat should be tilted backwards* so that the buttocks will not slide forwards. A tilt of 14−24° to the horizontal is recommended.
2. The backrest should be inclined at the following angles:
   to the seat   *105−110°*
   to the horizontal   *110−130°*
3. *The backrest should be provided with a lumbar pad*, as Akerblom said in 1948, when he pioneered the study of seating. The apex of this pad should meet the spine between the third and fifth lumbar vertebrae. This means that its vertical height above the back of the seat should be 100−180 mm. The pad should reduce kyphosis of the lumbar region, and hold the spine in as natural a position as possible.

The preferred shape of the multipurpose seat shown in Figure 52 is characterized by a slightly moulded seat surface in order to prevent the buttocks from sliding forward and by a high backrest with a lumbar pad fixed at a height between 100 and 200 mm above the seat.

The work seat

As far as work seats are concerned, our own researches indicate that a high backrest, slightly concave to the front at its top end, and distinctly convex in the lumbar region, is good both medically and ergonomically. Such a seat profile gives support to the lumbar region when the occupant is leaning forwards (in a working attitude) yet relaxes the back muscles thoroughly when leaning backwards, because it then holds the spine in a natural position.

Office chairs

Hünting and Grandjean [122] studied office chairs with high backrests under practical working conditions, recording sitting habits and reports of physical discomfort in different parts of the body. A tiltable chair and a similar model with a fixed seat were compared with a traditional type of chair fitted with an adjustable but short backrest. The subjects were doing their normal work while using each of the three chairs for 2 weeks. The most interesting results related to the reported preferences, as shown in Figure 53.

The survey indicated quite clearly that the office workers favoured the two types of chairs with the high backrest. This confirms the view expressed earlier that a high backrest is preferable for office work as most employees often desire to lean back. It is obvious that a high backrest will be more

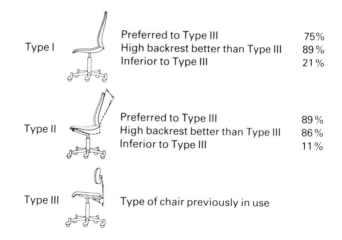

Type I		Preferred to Type III	75%
		High backrest better than Type III	89%
		Inferior to Type III	21%

Type II		Preferred to Type III	89%
		High backrest better than Type III	86%
		Inferior to Type III	11%

| Type III | | Type of chair previously in use | |

Figure 53. Comparative assessment of three experimental chairs, which were used by 66 office workers over a period of two weeks. *Type I= fixed moulded chair with high backrest. Type II=moulded chair which tilts 2° forwards and 14° backwards, freely movable with high backrest. Type III= standard office chair with adjustable backrest. After Hunting and Grandjean.* [122].

effective in supporting the weight of the trunk than a chair with a small backrest.

It follows that *any office working place offering the possibility of leaning back — all the time or only occasionally — should be provided with a chair with a high backrest.* It must be pointed out here that the experimental tiltable chair could not be fixed at the desired inclination; thus it did not provide enough support for the whole body. This was criticized by many subjects and leads to the conclusion that tiltable chairs or chairs with adjustable backrest inclinations should be fitted with a mechanism allowing the subject to fix the desired degree of inclination.

General experience as well as a number of studies have yielded the following 'golden rules' for office chairs:

Golden rules for office chair equipment

1. *Office chairs must be adapted to both the traditional office job and the modern equipment of information technology, especially to jobs at VDT workstations.*
2. *Office chairs must be conceived for a forward and reclined sitting posture.* See **Figure 54.**
3. *The backrest should have an adjustable inclination.* It should be possible to lock the inclination at any desired position.
4. *A backrest height of 480−520 mm vertically above the seat surface is a necessity.* The upper part of the backrest should be slightly concave. A breadth of 320−360 mm for the backrest is advisable. It may, as a further advantage, be concave in all horizontal planes with a radius of 400−500 mm.

Figure 54. An office chair must be conceived for a forward as well as a backward inclined sitting posture.
The lumbar spine must get proper support from the backrest in both sitting postures.

5. *The backrest must have a well-formed lumbar pad*, which should offer good support to the lumbar spine between the third vertebra and the sacrum, e.g., at a height of 100–200 mm above the lowest point of the seat surface. These recommendations are illustrated in Figure 55.
6. *The seat surface* should measure 400–450 mm across and 380–420 mm from back to front. A slight hollow in the seat, with the front edge turned upwards about 4–6° will

Tilting chair with high backrest

Backrest:

    Height (above seat) 50 cm

    lumbar pad

    slightly concave at thorax level

    adjustable inclination (104°–120°) with locking device

(don't forget a foot rest)

Figure 55. A well designed office chair should have a backrest according to the above mentioned recommendations.

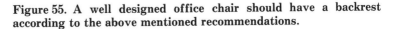

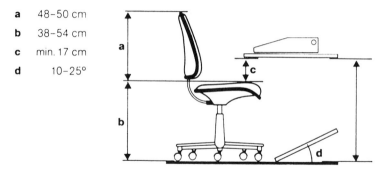

a	48–50 cm
b	38–54 cm
c	min. 17 cm
d	10–25°

Figure 56. Recommended dimensions for the design of the seat and working desk.

prevent the buttocks from sliding forwards. A light padding, rubber 20 mm thick, covered with non-slip, permeable material is a great aid to comfort.

7. *Foot rests are important*, so that short people can avoid sitting with hanging feet.

8. *An office chair must fulfil all the requirements of a modern seat*: adjustable height (380–540 mm), swivel, rounded front edge of the seat surface, castors or glides, 5-arm base and user-friendly controls. The most important dimensions of a seat and working desk are shown in Figure 56.

The backrest is crucial for an office chair

Orthopaedics as well as ergonomics recommend frequent or at least occasional changes of position from leaning forward to leaning back and vice versa. This calls for a 'dynamic' chair which allows easy changes of sitting posture. It is obvious that an adjustable backrest inclination is crucial for such a 'dynamic' chair. This can be achieved by having a tiltable shell or a backrest tilting independently of the seat surface which, on its part, can remain in a horizontal position or be tilted backwards with increasing backrest declination. With the tiltable shell the angle between backrest and seat surface remains the same in all positions. A drawback of this chair is the elevation of the knees with full backrest declination. With the independently tilting backrest a simultaneous inclining of the seat surface is advisable to prevent forward sliding.

The 'Syntop' chair

The most sophisticated office chair, especially adapted to VDT workstations, is the 'Syntop' model which was presented by Hort at the Ergodesign 84 conference in Montreux [119]. The development of this chair is based on the fact that with an ordinary chair the lumbar support of a backrest moves about 45 mm upwards when the inclination is increased from 90° to 105°. This corresponds to almost a whole lumbar vertebral segment. As a consequence of a declined backrest the lumbar support is no longer at the correct level but too high. The main

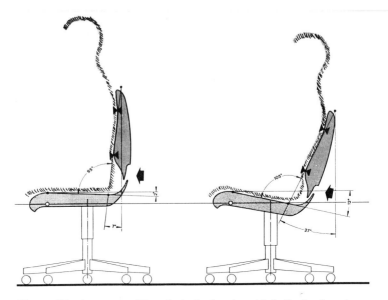

Figure 57. A recent office chair design in which the backrest moves down with increasing declination.
This mechanism allows the back to get adequate support at the correct level for any backrest declination, as indicated by the small arrows. The large arrows show the way the backrest descends with increasing declination.

characteristic of the new chair is that the backrest descends as its declination increases. This is shown in Figure 57.

The 'Syntop' chair model is an interesting example of applying ergonomics to the design of a chair.

The main objection to a good office chair with an adjustable backrest inclination is of course its cost. But one should bear in mind that the life span of a well-constructed chair is about 10 years, or about 2000 working days. The price of a good chair, which reduces physical discomfort and promotes well-being is, at a few pence per day, certainly a good investment!

5.5. The design of VDT workstations

The present metamorphosis of offices

At present VDTs are invading all types of offices. They are often entering a world where machines have not been used before. The result is a considerable change in offices and in the working conditions therein. To call the present change a metamorphosis, similar to that of caterpillars and butterflies, is hardly an exaggeration.

At the traditional office desk an employee performs a great variety of physical and mental activities and has a large space for various body postures and movements: he or she might look for documents, take notes, file correspondence, use the telephone, read a text, exchange information with colleagues, type for a while and he or she will leave the desk many times during the course of the working day. A desk which is too low or too high, an unfavourable chair, insufficient lighting conditions or other ergonomic shortcomings are not likely to cause annoyance or physical discomfort. The great choice of activities prevents adverse effects of long-lasting invariable physical or mental loads.

The situation is, however, entirely different for an operator working with a VDT for several hours without interruption or perhaps for a whole day. *Such a VDT operator is tied to a man—machine system.* His or her movements are restricted, attention is concentrated on the screen and the hands are linked to the keyboard. VDT operators are more vulnerable to ergonomic shortcomings: they are more susceptible to constrained postures, poor photometric display characteristics and inadequate lighting conditions. This is the reason why the computerized office calls for ergonomics; consequently the VDT workstation has become the launch vehicle for ergonomics in the office world.

Reports on discomfort

As long as engineers and other highly motivated experts operated VDTs, nobody complained about negative effects. However, the situation changed drastically with the expansion of VDTs into workplaces where traditional working methods had formerly been applied. Complaints from VDT operators about visual strain and physical discomfort in the neck/shoulder area and in the back became more and more frequent. This has provoked different reactions: some believe that the complaints are highly exaggerated and mainly a pretext for social and political claims, while others consider the complaints to be symptoms of a health hazard requiring immediate measures to protect operators from injuries to their health. *Ergonomics as a science stands between these opposing beliefs; its duty is to analyse the situation objectively and to deduce guidelines for the appropriate design of VDT workstations.*

Controversial field studies

Many field studies have been carried out into complaints about physical discomfort, often localized to the neck/shoulder/arm area. These studies have used self-rating questionnaires and included different types of VDT jobs as well as control groups. In most cases the non-VDT groups differ from the VDT groups not only in the use of VDTs but also in many other respects.

The problem of control groups is indeed very intricate. The introduction of VDTs is normally accompanied by changes in task design, speed of work and especially by major differences in performance and productivity. It is therefore not surprising that these field studies disclosed controversial results. If the control groups were engaged in traditional office work with low productivity and a great variety of activities they seem to be much less affected by musculoskeletal discomfort than the VDT groups. If, on the contrary, the control groups do strenuous work, like full-time typing, the complaints might be as frequent as among VDT groups. (An example of this phenomenon is shown in Figure 58.) All these field studies are fully described in the book *Ergonomics in Computerized Offices* [85]; here we shall discuss only a few studies related to the proper design of VDT workstations.

Medical findings   In this context an interesting contribution was made by Läubli *et al.* [124, 156], who applied investigation methods which are used in rheumatology and include the assessment of joint and back mobility, painful pressure points at tendons or other characteristic locations and painful reactions to muscle palpation. He noticed a high correlation between the medical findings of two examining doctors and between the medical findings and self-rated physical discomfort [159]. The study included 295 subjects engaged in two VDT jobs and in two non-VDT activities. Some of the results are presented in Table 8 and in Figure 58.

The results of Table 8 show that medical findings indicating musculoskeletal troubles in muscles, tendons and joints were frequent in the groups using data-entry terminals and among full-time typists, whereas the group performing traditional office work consisting of many different activities and movements showed the lowest figures. The palpation findings in the shoulders, listed in Figure 58, disclose a similar distribution of symptoms. Both the complaints and the medical findings must be taken seriously, especially since 13—27% of the examined employees had consulted a doctor for this reason.

Table 8. Incidence of medical findings in the neck—shoulder—arm area of office employees.
n = number of subjects.

Medical findings	Data-entry tasks (n = 53) (%)	Conversational tasks (n = 109) (%)	Full-time typists (n = 78) (%)	Traditional office jobs (n = 54) (%)
Tendomyotic pressure pains in shoulders and neck	38	28	35	11
Painfully limited head movability	30	26	37	10
Pains during isometric contractions of forearm	32	15	23	6

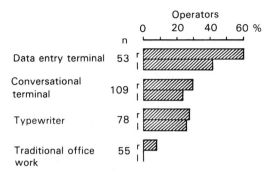

Figure 58. Palpation findings in the shoulders of four groups of office workers.
Painful pressure points are at the tendons, joints and muscles. r=right; l=left; n=number of examined operators. Differences between groups were significant at P < 0·01; Kruskal−Wallis test. According to Läubli [156].

Physical discomfort related to workstation design

In this field study on VDT operators [124,156,157] several significant relationships were discovered between the design of workstations or postures on the one hand and the incidence of complaints or medical findings on the other. These results can be summarized as follows:

Physical discomfort and/or the number of medical findings in the neck/shoulder/arm/hand area are likely to increase when:

The keyboard level above the floor is too low.
Forearms and wrists cannot rest on an adequate support.
The keyboard level above the desk is too high.
Operators have a marked head inclination.
Operators adopt a slanting position of the thighs under the table due to insufficient space for the legs. This is illustrated in Figure 59.

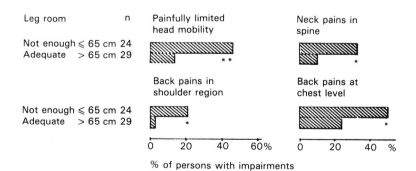

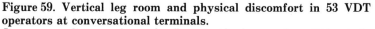

Figure 59. Vertical leg room and physical discomfort in 53 VDT operators at conversational terminals.
*Leg room = distance from the floor to the lower edge of the desk. n = number of operators. *P < 0·5, **P < 0·01, Mann−Whitney U test. According to Läubli and Grandjean [157].*

Operators disclose a marked lateral adduction (ulnar deviation) of the hands when operating the keyboard.

The frequent complaints about physical discomfort among VDT operators induced the office furniture manufacturers to place adjustable VDT workstations on the market. At the same time several experiments with adjustable workstations were carried out both under laboratory conditions and during practical office work. Since the latter gave important indications for the ergonomic design of VDT workstations they will be described below.

Preferred settings of VDT workstations in offices

A study on postures and preferred settings of adjustable VDT workstations during subjects' usual working activities was carried out by Grandjean *et al.* [94]. The experiments were conducted on 68 operators (48 females and 20 males aged 28 years on average) in four companies: 45 subjects had a conversational job in an airline company, 17 subjects had primarily data entry activities in two banks and 6 subjects were engaged in word processing operations. Each subject used the adjustable workstation shown in Figure 60 for one week.

The keyboard height was 80 mm above desk level. A chair was provided with a high backrest and an adjustable inclination. For the first two days a forearm/wrist support was used; on the following two days the subjects operated the keyboard without support and on the last day they were given the option to use it or not. Document holders were provided as an

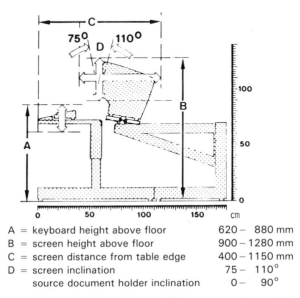

A = keyboard height above floor   620 – 880 mm
B = screen height above floor   900 – 1280 mm
C = screen distance from table edge   400 – 1150 mm
D = screen inclination   75 – 110°
    source document holder inclination   0 – 90°

Figure 60. The adjustable VDT workstation with the ranges of adjustability, used in a field study during subjects' usual working activities.

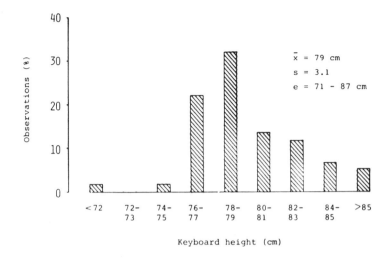

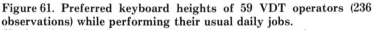

Keyboard height (cm)

Figure 61. Preferred keyboard heights of 59 VDT operators (236 observations) while performing their usual daily jobs.
Keyboard height=home row above floor; x̄=mean value; s=standard deviation; e=range.

optional supporting device for each subject. The preferred settings and postures were assessed and determined every day.

The analysis of the results of preferred settings disclosed no noticeable differences across the five days. In other words, the mean values remained practically the same for the whole week, independent of the use of wrist support. Thus the data obtained during the week could be put together for evaluation.

The frequency distribution of all preferred keyboard heights is reported in Figure 61.

Range of adjustability for keyboard desk

The 95% confidence interval lies between 730 and 850 mm. A desk level between 630 and 790 mm suits a keyboard of 80 mm, a level between 680 and 840 mm would be adequate for a keyboard of 30 mm. Assuming the 95% confidence interval, *the range for the adjustability of desk levels lies between 650 and 820 mm.* This seems to be a reasonable recommendation for workstation manufacturers.

The results obtained in this field study reveal slightly higher keyboard levels than those obtained in comparable laboratory studies. It is assumed that in short-term experiments subjects are less relaxed, sit more upright and try to keep the elbows low and the forearms in a horizontal position, thus giving preference to a slightly lower keyboard height.

All the results of preferred settings are assembled in Table 9.

Preferred settings of screen

The preferred screen heights and screen inclinations are in some cases influenced by the operators' attempts to reduce

Table 9. Preferred VDT workstation settings and eye levels during habitual working activities.

$n_1 =$ *number of subjects*; $n_2 =$ *number of observations. Visual down angle and screen inclination are related to a horizontal plane.*

Adjustable dimensions	n_1	n_2	mean	range
Seat height (mm)	58	232	480	430−570
Keyboard height above floor (mm)	59	236	790	710−870
Screen height above floor (mm)	59	236	1030	920−1160
Visual down angle, eye to screen centre (°)	56	224	$-9°$	$+2° - -2°$
Visual distance, eye to screen centre (mm)	59	236	760	610−930
Screen upward inclination (°)	59	236	$94°$	$88° - 103°$
Eye level above floor (mm)	65	65	1150	1070−1270

reflections. In fact, many operators reported less annoyance by reflections if they could adjust the screen.

The capital letters on the screen were 3·4 mm high corresponding to a comfortable visual distance of 680 mm. At the adjustable VDT workstation the operators tended to choose greater viewing distances; 75% of them had visual distances between 710 and 930 mm. No explanation can be found for this preference.

The visual down angles correspond well to the 'normal' line of sight, but they are not in agreement with those authors who recommend down angles of 38° or more [118, 163]. VDT operators obviously prefer slightly declined visual down angles of 0−15° (=90% confidence).

Correlations with anthropometric data

The calculation of Pearson correlation coefficients between anthropometric data and preferred settings revealed only poor relationships: between eye level and screen height $r = 0·25$ ($P = 0·03$) and between stature and keyboard height $r = 0·13$ (not significant). Some of the laboratory studies revealed similar results of poor or no relationships. It can be concluded, therefore, that *the preferred settings of VDT workstations are hardly influenced by anthropometric factors*; the main influence is individual habits.

Preferred postures

The most striking result of this field study concerns the postures associated with the preferred settings. The operators moved very seldom and did not noticeably change the main postural elements which are obviously determined by the position of keyboard and screen. Figure 62 shows the distribution of determined trunk postures expressed as angles of a line 'shoulder articulation to trochanter' related to a horizontal plane.

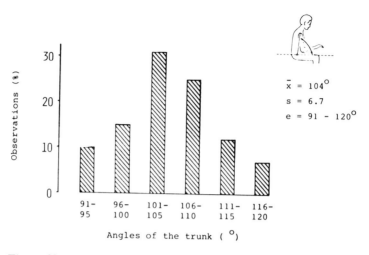

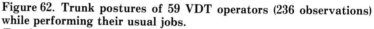

Figure 62. Trunk postures of 59 VDT operators (236 observations) while performing their usual jobs.
Trunk posture is assessed as the angle of a line between the hip and shoulder points to the horizontal. x̄ = mean value; s = standard deviation; e = range.

Most operators lean back...

The trunk inclinations approximate a normal distribution. The majority of subjects prefer trunk inclinations between 100 and 110°. Only 10% demonstrate an upright trunk posture. Figure 63 illustrates the mean and the range of observed trunk postures. It is obvious that the majority of operators lean back. This is the basis for all other adopted postural elements:

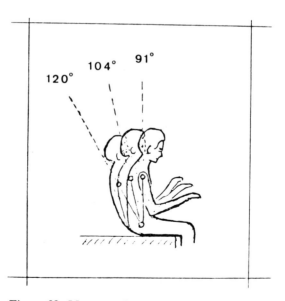

Figure 63. Mean and range of observed trunk postures of 59 operators.
Trunk posture is assessed as the angle of a line between the hip and shoulder joints to the horizontal.

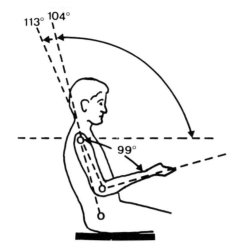

Figure 64. The 'average posture' of VDT operators at workstations with preferred settings.

the upper arms are kept higher and the elbow angles slightly opened. The mean figures for preferred trunk−arm positions are listed in Figure 64.

...and hold arms and hands slightly raised... It must be pointed out here that about 80% of the subjects do rest their forearms or wrists if a proper support is available. If no special support is provided, about 50% of the subjects rest forearms and wrists on the desk surface in front of the 80 mm high keyboard.

The observed postures are not due to the experimental workstation, for the measurements carried out at the previous workstations had already revealed nearly the same trunk and arm inclinations.

This study confirms a general impression one gets when observing the sitting posture of many VDT operators in

Table 10. Means($\bar{x}$), standard deviations (SD) and ranges of postural measurements obtained from VDT operators during their daily work at workstations with preferred settings.
59 operators, 236 observations.
[a] *Angle C7−earhole−vertical.*
[b] *See Figure 62.*
[c] *Abduction = lateral raising of the upper arm.*

Postural element	$\bar{x}$	SD	range
Trunk inclination (°)	104	6·7	91−120
Head inclination[a] (°)	51	6·1	34− 65
Upper arm flexion[b] (°)	113	10·4	91−140
Upper arm abduction[c] (°)	22	7·7	11− 44
Elbow angle (°)	99	12·3	75−125
Lateral abduction of hands (°)	9	5·5	0− 20
Acromion−home row distance (mm)	510	50	420−620

... like car drivers

offices: most of them lean back and often stretch out the legs. They seem to put up with having to bend the head forward and lift their arms. In fact, *many VDT operators in offices disclose postures very similar to those of car drivers*. This is understandable: who would like to adopt an upright trunk posture when driving a car for hours?

The results of all the measured postural elements, expressed as mean values and ranges, are reported in Table 10.

Preferred settings and physical discomfort

The VDT operators completed a questionnaire relating to feelings of relaxation and physical discomfort, once at the previous workstation and twice at the adjustable workstation with preferred settings. An index was calculated from the answers 'relaxed', 'tense' and 'impaired' for each of the involved parts of the body (neck, shoulders, back, forearms and wrists). In Figure 65 the mean indices of complaints from the previous workstation are compared with those reported on the second and on the fourth day. An index of less than $0 \cdot 5$ means that the majority of the subjects rated their postures as relaxed; an index of more than $0 \cdot 5$ means that many subjects indicated that their muscles were tense or even that they experienced impairments. From the figure it is obvious that

Discomfort is reduced with preferred settings

the indices were distinctly higher at the previous workstation than with the preferred settings. A chi-squared (χ^2) analysis showed significant differences between the previous and adjustable workstation for the neck, shoulders and back. It must be pointed out that at the previous workstation subjects sat on traditional office chairs with relatively small backrests.

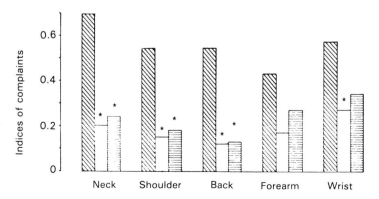

Figure 65. **Mean indices of complaints at the original workstation and the redesigned workstation with preferred settings.** *0 = relaxed; 1 = tense; 3 = impaired.* $^*P \leqslant 0 \cdot 05$.

At the adjustable workstations, however, they were provided with particularly suitable office chairs, featuring high backrests with adjustable inclinations, which allowed the whole back to relax. It is therefore reasonable to assume that the decrease in physical discomfort reported at the adjustable workstation was due to both the preferred settings and the proper chairs.

Confirmed results by the Shute and Starr study

These results were confirmed by Shute and Starr [241]. In a first field study telephone operators were provided with an adjustable VDT table and in a second study with an additional conveniently adjusted chair. Subjects used the adjustable workstation for several weeks while doing their normal work. The previous VDT table had a fixed height of 686 mm and the screen was 400 mm above the table. The previous chair was difficult to adjust and had an unsuitable backrest. The main difference to the new advanced chair was its easy adjustability. The results revealed a reduction of discomfort when either conventional component was replaced by an advanced component. But the reduction of physical discomfort was far greater when the advanced table was used together with the advanced chair. The authors concluded that the benefit of an advanced table can only be fully realized if it is used in combination with an advanced chair.

Preferred settings at CAD workstations

Van der Heiden and Krueger [107] examined the use and acceptance of an adjustable workstation for CAD operations. Height and inclination of the work surface as well as height, inclination, rotation and distance of the monitor could be adjusted with the aid of motorized devices. To study the use of the adjustment fixtures a continuous registration of settings was carried out during one week. The majority of operators had more than six weeks' experience in using the adjustable workstation. In the test week eight women and three men were studied during their normal CAD work, consisting of mechanical design. A total of 67 CAD work sessions were registered. Questionnaires and preferred settings were obtained from 11 female and 4 male operators. Of a total of 166 registered adjustments, 142 (=86%) were made at the beginning of a work session and 24 (=14%) were readjustments. Short operators used the adjustment device more frequently than tall people. Furthermore, operators who had not received specific instructions adjusted less frequently than others who had been given such instructions. The preferred settings are presented in Table 11.

The mean seat height of 540 mm is quite unusual. Another striking result is the forward tilting of the monitor with a preferred mean angle of $-8°$. The operators claimed that with this setting reflections resulting from windows behind them could be avoided. For that reason most operators preferred a

Table 11. Preferred settings of 15 operators at an adjustable CAD workstation.
$\bar{x}$ = *mean values*; *e* = *range.*
"Negative tilt = *a forward monitor inclination (top of the screen toward the operator).*

	$\bar{x}$	e
Seat height (mm)	540	500–570
Work surface height (mm)	730	700–800
Monitor centre above floor (mm)	1130	1070–1150
Monitor visual distance (mm)	700	590– 780
Work surface tilt ($^{\circ}$)	8·6	2–13
Monitor tilt^a ($^{\circ}$)	−7·7	−15−+1

relatively high monitor setting. All other preferred dimensions are similar to those of the VDT operators shown in Table 10.

Wishful thinking of standards versus operators' instinctive behaviour

Let us come back to the backward declination of the trunk observed at VDT workstations. This leaning back does not correspond at all with the commonly published and recommended values for postures [13, 36, 55, 264]. Figure 66 illustrates the great gap between 'wishful thinking' (recommendations) and actual postures. An important question suggests itself here. Is the upright posture healthy and therefore recommendable or is the relaxed position with the reclining trunk to be preferred? As already mentioned, by increasing the backrest declination from 90° to 120° a significant decrease in discal load and muscle strain is achieved [5], as shown in Figure 50. These orthopaedic studies [5] suggested *that*

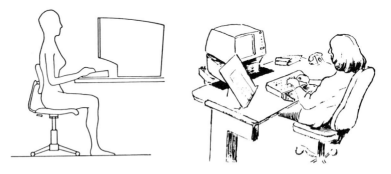

Wishful thinking   Preferred body posture

Figure 66. Recommended and actual postures at office VDT workstations.
Left: the upright trunk posture with elbows down and forearms almost horizontal, postulated in many brochures and standard works.
Right: the actual posture most commonly observed at VDT workstations resembles the posture of a car driver.

resting the back on a sloping backrest transfers a relevant portion of the trunk weight to the backrest and reduces strain on discs and muscles more than it does when sitting straight and upright. It is therefore concluded that VDT operators instinctively do the right thing when they prefer a reclined sitting posture and ignore the recommended upright trunk position.

One restriction must be made here: some special work situations (such as manual work requiring freedom of movement or physical effort) might call for an upright trunk position with elbows down and forearms in a horizontal plane. As already stated, the old mechanical typewriters requiring key forces of nearly 5 N were more easily operated in such a posture. But the advances in electronic keyboard technology today permit very rapid keying with low key forces of $0 \cdot 4 - 0 \cdot 8$ N and key displacements of $3 - 5$ mm. The new keyboard is mainly operated by finger movements with hardly any assistance by the forearms. These conditions might to some extent explain why VDT operators in offices prefer to lean back, keep the upper arms slightly forward with the wrists on a support (which can be the desk itself) and to adjust the keyboard to a rather high level.

Guidelines for the design of VDT workstations

From the study on preferred settings the following guidelines for the design of a VDT workstation can be proposed.

1. *The furniture should, in principle, be conceived as flexible as possible. A proper VDT workstation should be adjustable in the following dimensions:*

Keyboard height (floor to home row)	*700— 850 mm*
Screen centre above floor	*900—1150 mm*
Screen inclination to horizontal	*88°—105°*
Keyboard (home row) to table edge	*100— 260 mm*
Screen distance to table edge	*500— 750 mm*

2. *A VDT workstation without adjustable keyboard height and without adjustable height and distance of the screen is not suitable for a continuous job at a VDT.*
3. *The controls for adjusting the dimensions should be easy to handle, particularly at workstations with rotating shiftwork.*
4. *At knee level the distance between the front table edge and the back wall should not be less than 600 mm and at least 800 mm at the level of the feet.*

5.6. The design of keyboards

Parallel rows require unnatural hand positions

The keyboard for typing letters was invented in 1868. It was a mechanical device with four parallel rows of keys. To operate these keys rapidly, the typist must hold the hands parallel to

Figure 67. Position of wrists and hands operating a traditional keyboard.
The parallel position to the rows requires an inward rotation of the forearms and wrists and a sideward twisting (lateral abduction) of the hands.

the rows. This requires an unnatural position of the wrists and hands, characterized by an inward rotation of the forearms and wrists and a lateral (ulnar) adduction of the hands. These constrained postures often cause physical discomfort and in some cases even inflammations of tendons or tendon sheaths in the forearms of keyboard operators. Figure 67 illustrates such constrained postures of the wrists and hands at a keyboard.

With the development of electronics the mechanical typewriter was replaced by the electric. The mechanical resistance of keys was much reduced and the operation of the keyboard was made easier, but the unnatural position of wrists and hands remained.

Flat keyboards at VDTs

At VDT workstations the typing activity is similar to the traditional operation of typewriters. There are some slight differences, however. First, the number of keys has increased, with specially arranged numerical keys and several functional keys for operating the computer. Second, in conversational jobs operators must frequently wait for the response of the computer. This response time can last from one to several seconds. According to a Swedish study by Gunn Johansson and Gunnar Aronsson [135] response times of more than 5 s were experienced as annoying and stressing. During these unwanted pauses operators like to rest forearms and wrists on suitable supports. This induced some VDT designers to develop flat keyboards which allow operators to rest their forearms and wrists on the desk. For this reason many ergonomists nowadays recommend *a flat keyboard with a home row not higher than 30 mm above the desk and the possibility for the operator to shift the keyboard on the desk according to needs.*

The next step should be an ergonomic design of the keyboard in order to avoid the constrained hand postures by reducing or eliminating the inward rotation and lateral adduction of the hands and by a proper support for forearms and wrists.

Studies on split keyboards

Studies along this line were carried out in 1926 by Klockenberg [143] and in 1965 by Kroemer [147,150] who proposed splitting the keyboard into two parts and arranging them in such a way that the hands could be kept in a more natural position. Kroemer's halved keyboards had an opening angle of 30° and a lateral declination adjustable between 0 and 90°. The opening angle is defined by two lines running through the inner board of the keys Y-H-N and T-G-B. Experiments disclosed that the experimental keyboard generated less painful fatigue than traditional typewriters.

EMG of forearms and shoulders

Recently, Zipp *et al.* [285] studied the electrical activity of various muscles in the shoulder/arm area in relation to hand/arm postures according to the characteristics of keyboard operations. With increasing lateral adduction of the hands from a neutral position, an increase in the electrical activity of the muscles involved was recorded. A split keyboard, as proposed by Kroemer, disclosed decreased electrical activities in the arm/shoulder area, obtained already with lateral keyboard inclinations of 10−30°. The same result was observed when the angle between the two keyboard halves was opened. The authors concluded that the static muscle load in the arm/shoulder area is significantly reduced with such a keyboard design.

Experiments with split keyboards

Following this line of research, Grandjean *et al.* [93], Hünting *et al.* [125] and Nakaseko *et al.* [197] developed an adjustable model of a split keyboard and studied the preferred settings of opening angles, lateral sloping and distances between the split keyboards on 51 subjects. Typing with the split keyboard with preferred settings decreased the lateral adduction of the hands shown in Figure 68, reduced discomfort and increased feelings of being relaxed in the neck/shoulder/arm/hand area.

Effects of a large forearm/wrist support

The use of a large forearm/wrist support was associated with a declined sitting posture of the subjects and with an increased pressure load of the forearm/wrist on the support, reaching mean values of nearly 40 N (4 kg). Such weight transfers onto

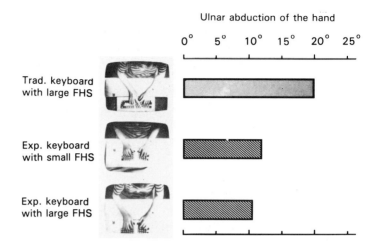

Figure 68. Mean angles of sideward twisting (lateral abduction) of the right hand with three types of keyboards.
Top: traditional typewriter with large forearm-wrist support (200 mm).
Middle: split keyboard with an opening angle of 25° and a lateral slope of 10°. Small forearm-wrist support (100 mm).
Bottom: the same split keyboard but with a large forearm-wrist support (200 mm).
FHS: forearm-hand support.

the support will strongly decrease the load on the intervertebral discs. Of 51 subjects, 40 preferred the split keyboard with the following characteristics:

Preferred settings of split keyboards

Angle between the two half-keyboards	25°
Distance between the two half-keyboards (measured as distance between the keys G and H)	95 mm
Lateral sloping of both half-keyboards	10°
A hand-configurated design of the keys	

A prototype of such a keyboard is shown in Figure 69. A commercial model was presented at the Ergodesign 84 conference by Buesen [35].

Guidelines for the design of VDT keyboards

In the last few decades, the manufacturers of typewriters have greatly improved the design and the mechanical characteristics of keys. There are almost no controversial opinions and the following guidelines are today widely accepted for the design of VDT keyboards:

Keyboard height above desk (middle row)		30 mm
Keyboard height (front side)	less than	20 mm
Inclination		5−15°
Distance between key tops		17−19 mm
Resistance of keys		0·4−0·8 N
Key displacement		3−5 mm

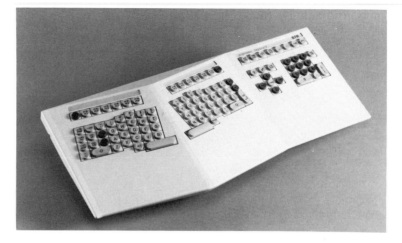

Figure 69. A keyboard designed in accordance with ergonomic principles.
The two keyboard halves show an opening angle of 25° in order to avoid a sideward twisting of the hands and are provided with lateral slopes of 10° to lessen the extent of inward rotation of the forearms and wrists. According to Nakaseko et al. [197].

The operator should feel when the stroke has been accepted; this is called tactile feedback. The best feedback quality is achieved when the point of acceptance and pressure is located about halfway down the key displacement.

Finally, the reader is once again reminded that the keyboard should be movable on the desk. A good support for forearms and wrists with a depth of at least 150 mm should be provided.

6. *Heavy work*

6.1. Physiological principles

Heavy work is any activity that calls for great physical exertion, and is characterized by a high energy consumption and severe stresses on the heart and lungs. Energy consumption and cardiac capacity set limits to the performance of heavy work, and these two functions are often used to assess the degree of severity of a physical task.

Mechanization has reduced the demands for strength and energy of the operator; nevertheless in many industries there are still jobs which rank as heavy work, and which not infrequently lead to overstrain. Thus, for example, Hettinger [113] reports that studies that he made between 1961 and 1969 in the German iron and steel industry showed excessive physical demands being made on the operators at about one-quarter of the 337 work places he investigated. Heavy work is also common in mining, building, haulage, agriculture and forestry; furthermore, today it is a major problem for ergonomics in developing countries.

Metabolism A fundamental biological process is to take in nutrients and to convert their chemical energy into mechanical energy and heat. Food is progressively broken down in the intestines until its constituents can pass through the gut wall and be absorbed into the blood. Most of the nutrients then pass to the liver, where they are stored as glycogen, an energy reserve. When needed, they again pass into the blood stream as readily usable compounds, mainly sugars. Only a small proportion of the food is used for building up body tissues, or reaches the adipose tissues as fat.

The blood carries nutrients to all the cells of the body, where they are broken down still further by precisely controlled stages, ending up as water, carbon dioxide and urea. The collective term for these processes is *metabolism*, which can be compared to a slow self-regulated combustion. The comparison is the more apt, since metabolism, like combustion, needs a supply of oxygen, which it obtains via the lungs and the blood stream. These metabolic processes liberate heat, as

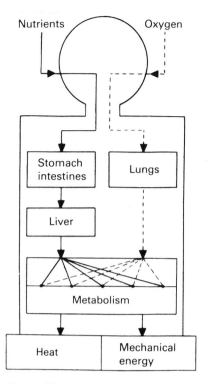

Figure 70. Diagram illustrating the conversion of nutrients into heat and mechanical energy in the human body.

well as mechanical energy, depending on what muscular activity is going on. Figure 70 shows these processes diagrammatically.

Energy consumption

Energy consumption in man is measured in kilojoules (kJ)†; it can be measured indirectly by measuring the consumption of oxygen since the two are directly related. *When one litre of oxygen is consumed in the human body, there is, on average, a turnover of 20 kJ of energy. This is called the calorific value of oxygen; to obtain the energy expenditure the consumption of oxygen (in l) must be multiplied by 20.*

Basal metabolism

Measurements show that a resting person has a steady consumption of energy, depending on size, weight and sex. When the person is lying down, with the stomach empty, this quantity is known as the *basal metabolism*. For a man weighing 70 kg it amounts to about 7000 kJ per 24 hours, and for a woman weighing 60 kg, about 5900 kJ. Under these conditions of basal metabolism nearly all the chemical energy from nutrients is converted into heat.

---

† The former measure was the kilocalorie, 1 kcal being 4·187 kJ

6.2. Energy consumption at work

As soon as physical work is performed, energy consumption rises sharply. The greater the demands made on the muscles by one's occupation, the more the energy consumed.

Work joules

The increased consumption associated with a particular activity is expressed in *work joules* and is obtained by measuring the energy consumption while working and subtracting from this the resting consumption, or basal metabolism.

Work joules indicate the level of bodily stress, and in relation to heavy work they can be used to assess the level of effort, to work out necessary rest periods, and to compare the efficiency of different tools and different ways of arranging the work. In this context it should be clearly understood that energy consumption measures only the level of physical effort; it tells us nothing about mental stress, about demands the work may make in alertness, concentration or skill, nor about any special physical problems such as excessive heat or static loads from awkward postures.

Hence energy consumption should be used as a measure of comparison only for strenuous physical effort and never for studying mental activities or skilled work.

Leisure joules

Everyday activities also consume energy, which we may call *leisure joules*. A fair average consumption would be 2400 kJ daily for a man and 2000−2200 kJ for a woman.

Thus the total energy expenditure is made up as follows:

1. Basal metabolism.
2. Work joules.
3. Leisure joules.

Energy expenditure at work

During and just after World War II a number of physiologists systematically studied the energy expenditure in a variety of occupations. At that time work joules were used as the basis

Table 12. **Energy demands of some occupational activities.**
The figures of kJ/day are approximate annual means of daily energy expenditure. After Lehmann [162].

Type of work	Example of occupation	kJ/day Men	kJ/day Women
Light work, sitting	Book keeper	9 600	8 400
Heavy manual work	Tractor driver	12 500	9 800
Moderate bodily work	Butcher	15 000	12 000
Heavily bodily work	Shunter	16 500	13 500
Extreme bodily effort	Coal miner Lumberjack	19 000	−

Table 13. Energy expenditure in work joules during various forms of physical activities.
The figures of kJ/min refer to the net working time. After Lehmann [162].

Activity	Conditions of work	kJ/min
Walking	Level, smooth surface, 4 km/h	8·8
Walking with load	30 kg load, 4 km/h	22·3
Climbing stairs	30° gradient, 17·2 m/min	57·5
Cycling	Speed 16 km/h	22·0
Sawing wood	60 double strokes/min	38·0
Household work	Cleaning, ironing	
	washing floors	8−20

for assessing occupational loads and the severity of tasks. Today these procedures are no longer in vogue and new methods are used to assess work loads. Nevertheless, some of these results are assembled in Tables 12 and 13.

Working posture might have a significant influence on the energy expenditure, as the example in Figure 71 shows.

Sitting   Standing   Stooping   Kneeling
3–5 %   8–10 %   50–60 %   30–40 %

Figure 71. Percentage increase in energy consumption for different bodily postures.
100% = energy consumption lying down. The relative increase, as a percentage of this, is the same for men and women.

Energy consumption and health

Most workers in industrial countries sit down at their work. If we add to this the time they sit while they travel to and from work, and in front of the television screen in the evening, *20th century man is clearly on the way to becoming a sedentary animal.* Such a sedentary life leaves many organs of the body under-used. Often more chemical energy is taken into the body than is consumed, leading to overweight, with increased risk of cardiac and circulatory disease, as well as metabolic troubles. Research has shown that a healthy occupation should involve a daily energy consumption of 12 000−15 000 kJ

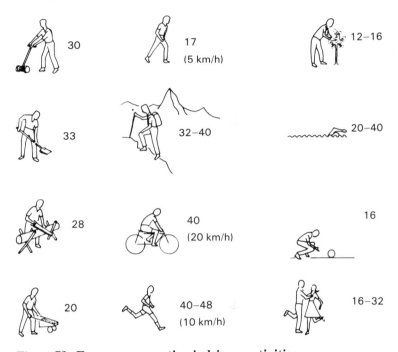

Figure 72. Energy consumption in leisure activities.
The figures are the average overall consumption in kJ/min, for men. Slightly lower figures apply to women, about 10—20% less. Drawn from figures given by Durnin and Passmore [63].

for a man and 10 000—12 000 kJ for a woman. This category includes, for example, being a non-motorized postman or a mechanic, a shoemaker, or having one of many jobs in building or agriculture. People who have sedentary jobs can make up some of the deficiency during their leisure time: some examples are given in Figure 72.

Upper limits of heavy work

Most work physiologists today consider an energy consumption of 20 000 kJ per working day (averaged over a year) to be a reasonable maximum for heavy work.

This corresponds to an average of 10 500 occupational kJ per working day, and if these are spread over an 8-hour work period they amount to 1300 kJ/h.

Seasonal workers may exceed these values for a few weeks, or even a few months, provided slack periods intervene.

This is true in principle for many heavy workers in forestry, haulage and so on, who can reach levels of 22 000—30 000 kJ for a few days without ill effects. This limit of 20 000 kJ per working day is therefore a yearly average which applies only to a heavy worker in a good state of health. Clearly there are individual variations both above and below this mean value, depending particularly upon such factors as constitution, level of training, age and sex.

6.3. Efficiency of heavy work

Heavy work is becoming increasingly rare in industrial countries, but it is still widespread in developing countries. A major target of ergonomics in developing countries is therefore the achievement of a high level of efficiency in heavy work.

Efficiency   As far as energy is concerned, a man doing physical work can be compared to an engine. An engine converts the chemical energy of coal or oil into mechanical performance, but with certain losses. Similarly, in the human body, a large part of the energy it receives is wasted by being converted into heat, and only a small proportion becomes useful mechanical energy. In both engine and man the term *efficiency* denotes the ratio between the externally measurable useful effort and the energy consumption that was necessary to produce it.

Under favourable conditions, human physical effort can be 30% efficient, turning 30% of the energy consumed into mechanical work and the remaining 70% into heat. But heat is not the only form in which energy is 'wasted'. Unproductive static effort is equally important. Hence the highest level of efficiency is possible only by converting as much as possible of the mechanical effort into a useful form, with little or none of it being dissipated in holding or supporting things. *The greater the proportion of mechanical energy to go into static effort,*

Table 14. **Maximum efficiency in various physical tasks.**

$$\% \ efficiency = \frac{useful \ work}{energy \ consumption} \times 100.$$

Activity	% efficiency
Shovelling in stooped posture	3
Screw driving	5
Shovelling in normal posture	6
Lifting weights	9
Turning a handwheel	13
Using a heavy hammer	15
Carrying a load on the back on the level, returning without load	17
Carrying a load on the back up an incline, returning without load	20
Going up and down ladders, with and without load	19
Turning a handle or crank	21
Going up and down stairs, without load	23
Pulling a cart	24
Cycling	25
Pushing a cart	27
Walking on the level, without load	27
Walking uphill on a $5°$ slope, without load	30

the lower the efficiency. This is particularly true when work is carried on with a bent back.

In every kind of heavy work it is important to aim at maximum physiological efficiency, not only to use energy economically, but to minimize stresses upon the operator. For this reason work physiologists have made a great many attempts to measure the physiological efficiency of various kinds of working methods, and the use of different tools and other equipment. Their results make it possible to formulate guidelines for the layout of work and the design of equipment, which are of particular importance where strenuous work is concerned.

Some examples are briefly enumerated in Table 14.

Shovelling

Shovelling is a common form of manual work, which has been thoroughly studied at the Max-Planck-Institut für Arbeitsphysiologie in Dortmund, Germany [162]. The highest level of efficiency was attained when *a load of 8—10 kg was shovelled 12—15 times per minute.* Besides the load, one must take into account the weight of the shovel itself, so use a big shovel when the material is light, and a small one for heavy materials. For fine-grained materials the shovel should be slightly hollow, spoon-shaped, and with a pointed tip for penetration. For coarse materials, the cutting edge should be straight, and the blade flat, with a rim round the back and sides. For stiff materials such as clay, the cutting edge can be either straight or pointed, but the blade should be flat. The handles of spades and shovels should be 600—650 mm long.

Sawing

In one of our own studies [87], we investigated the physiological efficiency of various types of timber saw. Oxygen consumption was measured before and during the use of five different types of saw and the number of work joules calculated. This energy consumption related to cutting slices or discs one square metre in area, and the results are shown in Figure 73.

The results show that the best saws were the two ripsaws with a broad blade and either two or four cutting teeth in each group, and these were preferred to the others. A more extensive study showed that the best results were obtained from a sawing rate of 42 double strokes per minute, and a vertical pressure of about 100 N.

Hoeing

Figure 74 shows the results of loosening the soil of a vegetable plot, using hoes of two different types. In soft soil a swivel hoe is much more efficient than the ordinary chopping hoe, but if the soil is hard and dry, the two types are about equally good.

Walking

A pleasant and not too strenuous walking pace is 75—110 steps per minute, with a length of pace between 0·5 and

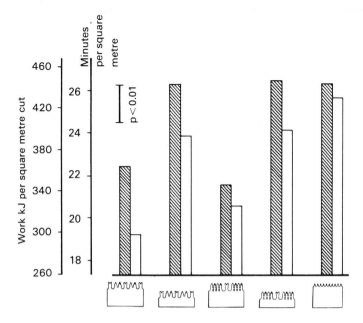
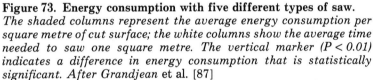

Figure 73. Energy consumption with five different types of saw.
The shaded columns represent the average energy consumption per square metre of cut surface; the white columns show the average time needed to saw one square metre. The vertical marker (P < 0.01) indicates a difference in energy consumption that is statistically significant. After Grandjean et al. *[87]*

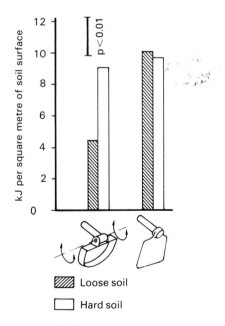

Figure 74. Energy consumption while using a swivel hoe (left) and a traditional hoe (right), in loose and hard soil.
After Egli et al. *[64].*

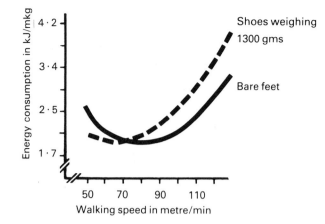

Figure 75. Best walking speed.
Ordinate: efficiency expressed as the energy consumption in kJ per unit of walking effort (kilogram metre). Curves shown for bare feet (solid line) and for shoes weighing 1300 grams (broken line). The heavier the shoes, the lower the best speed.

0·75 m, but this is not the most efficient pace when the work performed is compared with the energy consumed. Figure 75 shows the efficiency of walking, expressed in terms of the energy consumed per kgm of work performed.

From studies by Hettinger and Müller [114] it appears that the *most efficient walking speed is 4–5 km/h, reduced to 3–4 km/h in heavy shoes.*

Carrying loads

Work physiologists have given special attention to all kinds of jobs which involve carrying heavy loads, since these are rightly considered to be among the most strenuous forms of work.

According to Lehmann [162], 50–60 kg is the most efficient load to carry. Smaller ones are more convenient, but require more journeys, and 'carrying one's own body weight to and fro' increases the total consumption of energy. If we disregard the 'way back', then maximum efficiency, according to Teeple [253] is given by:

load: 35% of body weight
speed: 4·5–5 km/h.

Nature of the surface

The nature of the surface has a great effect on energy consumption. The following are the distances that can be covered with a load of one tonne on a 4-wheeled truck to use up 1050 work kJ [163]:

On a light railway track:	850 m
On a good road:	700 m
On a bad road:	400 m
On a dirt road:	150 m

In this connection it may be said that pushing the truck in front of one requires about 15% less effort than dragging it behind. The handles of the truck should be one metre above the ground and about 40 mm thick. A two-wheeled barrow should have its centre of gravity as low and as close to the axle as possible, for better balance and less weight to support.

Gradients

When a task involves climbing, a gradient of $10°$ gives the best efficiency. The following heights can be reached for an outlay of 42 work KJ:

On a ladder at $90°$:	$11·5$ m
On a ladder at $70°$:	$14·4$ m
On a staircase at $30°$:	$13·2$ m
On a path at $25°$	$13·1$ m
On a path at $10°$	$15·5$ m

Climbing ladders is most efficient when the ladder is at an angle of $70°$ and the rungs are 260 mm apart, but if heavy loads are being carried, 170 mm between rungs is better. If loads are often lifted to considerable height, some sort of lifting gear should be provided. Even a hand-operated crank halves the effort.

Staircases

Climbing stairs is one of the best forms of exercise in everyday life. *From the preventive medicine standpoint one is strongly advised to avoid using lifts whenever possible and to take every opportunity to climb the stairs as a 'gymnastic performance'.* At the same time it is sensible to have the stairs designed in such a way that they can be climbed as efficiently as possible. This is particularly so when the stairs are in constant use, or used by infirm and old people.

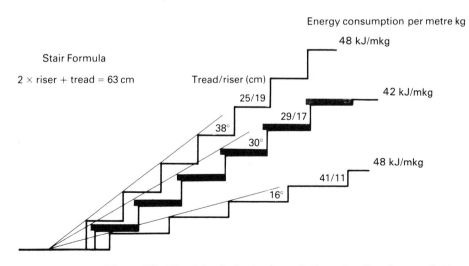

Energy consumption per metre kg

48 kJ/mkg

Stair Formula

2 × riser + tread = 63 cm

Tread/riser (cm)

25/19

42 kJ/mkg

29/17

38°

30°

48 kJ/mkg

41/11

16°

Figure 76. Physiological recommendations for the slope and dimensions of stairs.

Lehmann [162] found that least energy was consumed when climbing stairs with a gradient of 25−30°, and he fixed the following empirical norms:

(a) *tread height (riser):*   *170 mm*
(b) *tread depth:*   *290 mm*

Stairs of these dimensions are not only the most efficient, but also seem to cause the fewest accidents. This recommendation can be expressed as a formula:

$$2h + d = 630 \ mm$$

where h = height of riser and d = depth of tread, both in mm. Figure 76 gives optimum dimensions for the design of flights of stairs.

6.4. Heart rate as a measure of work load

Energy consumption and heart rate

Until 20 years ago energy consumption was usually the means by which the severity of physical stress was estimated, but now it is becoming more and more evident that energy consumption alone is not enough. The degree of physical stress depends not only on the number of kJ consumed, but also on the number of muscles involved and on the extent to which they are under static load. A given level of energy consumption is much more strenuous if it is achieved by using only a few muscles than if many others are employed. Similarly, the same energy consumption by static muscular effort is distinctly more tiring than if it is applied to dynamic work.

A further argument against the use of energy consumption as a measure of work load is the heat that often prevails. This

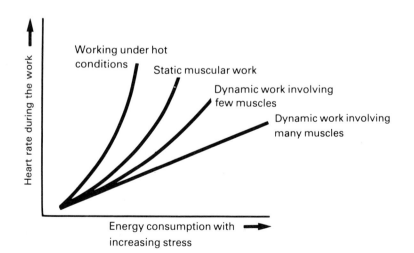

Figure 77. **Increase in heart rate associated with various types of stress.**

may be a trivial part of the energy consumption, yet may cause a sharp rise in the heart rate.

The various ways in which a rise in heart rate is related to work load are shown diagrammatically in Figure 77.

This diagram shows that a given energy consumption can make different demands on the heart according to the circumstances.

Summarizing, it can be said *the rise in heart rate with increasing work load is the steeper*:

(a) *the higher the ambient temperature*;
(b) *the greater the proportion of static to dynamic effort*;
(c) *the smaller the number of muscles involved*.

For these reasons, heart rate has been used more and more in recent years as an index of work load. Before going into details about heart rate and pulse we must first consider the relationships between the blood circulation and respiration.

Blood and respiration

Physical work demands adjustments and adaptations which affect nearly all the organs, tissues and fluids of the body. The most important adjustments are:

1. Deeper and more rapid breathing.
2. Increased heart rate, accompanied by an initial rise in cardiac capacity and an increased output per minute.
3. Vasomotor adaptations, with dilatation of the blood vessels in the organs involved (muscles and heart), while other blood vessels contract. This diverts blood from the organs not immediately concerned into those which need more oxygen and nutrients.
4. Rise in blood pressure, increasing the pressure gradient from the main arteries into the dilated vessels of the working organs, and so speeding up the flow of blood.
5. Increased supply of sugar, by release of more sugar into the blood from the liver.
6. Rise in body temperature and increased metabolism. The rise in temperature speeds up the chemical reactions of metabolism, and ensures that more chemical energy is converted into mechanical energy (for this reason athletes 'warm up' before a contest).

As work continues, secondary metabolic effects arise, particularly in the chemical composition of the body fluids. There is an accumulation of metabolic waste products, notably lactic acid, and the kidneys have more waste products to excrete. Muscular activity generates more heat in the interior of the body and to restore the balance more heat must be lost through the skin by increased blood flow and by sweating.

Within certain limits, some of these changes — ventilation of the lungs, heart rate and body temperature — show a linear relationship with the rate of energy consumption, or the work

Table 15. Metabolism, respiration, temperature and heart rate as indications of work load.

Assessment of work load	Oxygen consumption (l/min)	Lung ventilation (l/min)	Rectal temperature (°C)	Heart rate (pulses/min)
"Very low" (resting)	0·25−0·3	6−7	37·5	60−70
"Low"	0·5−1·0	11−20	37·5	75−100
"Moderate"	1·0−1·5	20−31	37·5−38·0	100−125
"High"	1·5−2·0	31−43	38·0−38·5	125−150
"Very high"	2·0−2·5	43−56	38·5−39·0	150−175
"Extremely high" (e.g., sport)	2·4−4·0	60−100	over 39	over 175

performed. Since these changes can be measured while a person is at work, they can be used to assess the physical effort involved. Table 15 shows reactions measured at various work loads, after Christensen [45].

Measuring the heart rate (pulse)

Measuring the heart rate ('taking the pulse') is one of the most useful ways of assessing the work load, because it can be done so easily.

One way is by simply feeling the pulses of the radial artery in the wrist, but to do this while a person is at work is an interruption and may disturb their work, thereby producing a false result. Instruments have been devised which give a continuous record of heart rate, and these have given consistent results during work. The best known of modern methods is the trace shown on an electrocardiogram. This records the action current in the heart muscle, and the heart rate is the number of R peaks (the strongest action potential) per unit of time.

Heart rate during physical activity

As already mentioned, and within certain limits, the heart rate increases linearly with the work performed, provided that this is dynamic not static, and is performed with a steady rhythm, only the force exerted being variable.

When the work is comparatively light, the heart rate increases quickly to a level appropriate to the effort, then remains constant for the duration of the work. When work ceases, the pulse returns to normal after a few minutes.

With more strenuous work, however, the heart rate goes on increasing until either the work is interrupted, or the operator is forced to stop from exhaustion. Figure 78 shows diagrammatically the behaviour of the pulse during certain work studies.

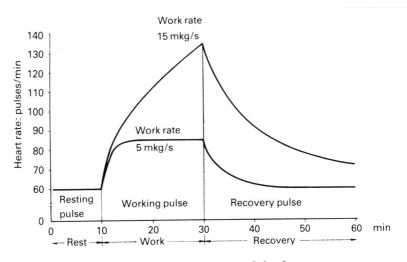

Figure 78. Heart rate for two different work loads.
At the higher level of stress the heart rate goes on increasing as long as the stress lasts, whereas at the lower rate it levels off at a 'plateau', or 'steady state'.

Scales of heart rate

Müller [190] proposed the following definitions:

Resting pulse: average heart rate before the work begins.
Working pulse: average heart rate during the work.
Work pulse: difference between the resting and working pulses.
Total recovery pulse (recovery cost): sum of heart beats from the cessation of work until the pulse returns to its resting level.
Total work pulse (cardiac cost): sum of heart beats from the start of the work until resting level is restored.

Müller [190] considers that the total recovery pulse is a way of measuring fatigue and recovery. Since 'fatigue' is a subjective term it would be more correct to regard heart rate, and particularly total recovery pulse, as measuring the physical work load of the individual.

Acceptable limits

Karrasch and Müller [140] made use of their studies to lay down an acceptable upper limit of work load as being that within which the working pulse did not continue to rise indefinitely, and when the work stopped, returned to the resting level after about 15 minutes. This limit seems to ensure that energy is being used up at the same rate as it is being replaced, that is to maintain a steady state. *The maximum output under these conditions is the limit of continuous performance throughout an 8-hour working day.*
The limit of continuous performance for men is reached when the average working pulse is 30 beats/min above the resting pulse (i.e., work pulse = 30/min), both of these being

measured in the same posture (e.g., both standing up), so that the static loads are the same. Rohmert and Hettinger [219] have made a systematic study of limits of work load, during which the heart rate remained steady, using a bicycle ergometer and a hand-crank for 8 hours at a time. They came to the conclusion that this limit was still valid up to a work pulse of 40/min, provided that *the resting pulse was measured when the operator was lying down*. The authors show that for dynamic work involving a moderate number of muscles 4 work kJ/min = 10 work pulses.

Several studies undertaken in factories have shown us that it is easier to measure the resting pulse when the subject is sitting than when lying down, so we suggest that, *for men*, we should start from a *resting pulse taken when seated*, and fix *35 work pulses as the limit for continuous performance*. There are no corresponding studies of *women*, but on physiological grounds it seems reasonable to postulate *30 work pulses as the limit for continuous performance*, again taking the resting pulse in a seated position.

Recovery pulse

In the USA Brouha [31] has made lengthy and detailed studies of heart rate as an index of work load, and the most important of his findings include:

1. *A performance of 360 mkg/min produced work pulses of 50 for women and 40 for men.*
2. The total recovery pulse (recovery cost) was as good as, or even better than, the total work pulse (cardiac cost) as a measure of work load.
3. The following procedure seems to be suitable for recording the recovery process: after work has stopped, take the pulse at the wrist during the following 30-second intervals:
   from 30 s to 1 min;
   from $1\frac{1}{2}$ min to 2 min;
   from $2\frac{1}{2}$ min to 3 min.
   Take the average of these three readings as being the heart rate during the recovery phase as well as an indication of the preceding work load.
4. Brouha recommended the following criteria for determination of acceptable limits of work load: *the first reading should not exceed 110 pulses/min with a fall of at least 10 pulses between the first and third readings*. Given these conditions the work load could be sustained during an 8-hour working day.

Combined effects of work and heat

As we have already seen, heart rate can be used as a measure of heat load as well as of work load. This is understandable, since the 'heart' is merely a 'pump' which supplies the muscles with blood on the one hand, and suffuses the skin with blood on the other, when it is necessary to eliminate excess heat. So when work is being performed under hot conditions the heart and circulatory system have two functions:

(*a*) *Transporting energy to the muscles.*
(*b*) *Transporting heat from the interior of the body to the skin.*

This double burden on the heart and circulatory system is a common occurrence in industry and agriculture. When heavy work has to be carried out in ambient temperatures† of 25°C, the elimination of excess heat throws an additional load upon the heart. An example of such a double loading is drop forging. Hünting and his colleagues [123] were able to show that holding the forging during the operation caused considerable loads on the back and arms, both static and dynamic. Working conditions were certainly strenuous, and deliberate pauses of up to 60% of the working time were noted. At the same time the operator was subjected to intensive radiation of heat. Figure 79 gives a simple sketch of the drop forge, and results from measurements of the work load on a 47-year-old operator are shown in Figure 80.

The mean work pulse of 41 beats/min during forging rates as a high level. Figure 80 also indicates the limit of continuous performance set by Müller [190], which is valid in this case because the resting pulse of the operator was measured when he was standing up.

The net working time is related to the number of forgings/min made and to the ambient temperature, and the heart rate follows in step with these two quantities.

Static work and heart rate

Chapter 1 gave an example of 'potato planting', where the 'static effort' produced a work pulse of 40 beats/min, which fell to 31 beats/min when the need to support the potato basket was eliminated (see Figure 8), although the energy consumption remained the same. This shows that *static effort can cause a rise in heart rate, even though there is no increase in the total consumption of energy. This rise must be interpreted as an increased physical stress.* This phenomenon is also shown in Figure 77.

In the laboratory, the effects of static muscular effort have been studied mainly by using loads which had to be either supported or dragged along. Figure 81 shows the results of experiments by Lind and McNicol [168]. It was found that in spite of considerable static loads the heart rate rose only a little above 100 beats/min, and that the initial resting pulse was quickly restored afterwards.

---

† Ambient temperature is a rough approximation of the mean between the air temperature and the heat radiated from the forge, and is a suitable way of assessing the heat load.

Figure 79. Drop-forging.
The operator manipulates a piece of glowing metal weighing 18·5 kg with tongs, and is exposed to a strong radiant heat.

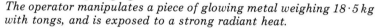

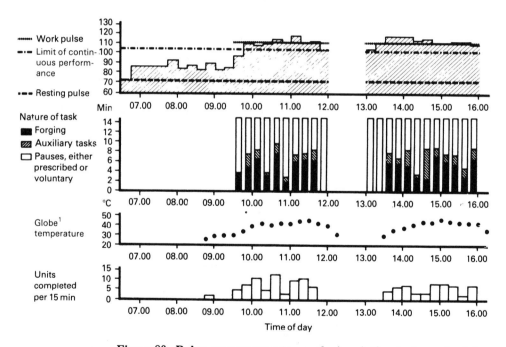

Figure 80. Pulse measurements, analysis of the task, and globe temperatures of an operator of a drop forge.
Top: heart rate (averaged over 15 minutes). Average work pulse during drop forging is 41/min. Middle: distribution in time of the various phases of the job. Pure drop forging occupies 28% of the total time. Globe temperatures for each phase are given in °C. Bottom: number of completed units per 15 minute period. After Hunting et al. [123].
[1]Globe temperature measures the Mean Radiant Temperature (see page 322).

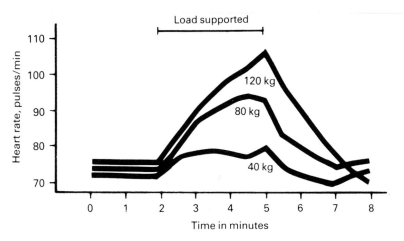

Figure 81. **Heart rate during static effort (carrying a load).**
The heavier the load, the greater the total of both work pulses and recovery pulses. After Lind and McNicol [168].

6.5. Case histories involving heavy work

Heavy work in the iron and steel industry

Hettinger [113] and Scholz [236] carried out intensive studies of work loads in the German iron and steel industry. In Figure 82 the maximum heart rate of 380 workers, measured over periods of 2 and 4 minutes, is displayed as a frequency distribution curve. It is evident from this that the most frequent peak value lies in the range 130–140 beats/min (mean 132·6), and that extreme increases of up to 180 beats/min occur.

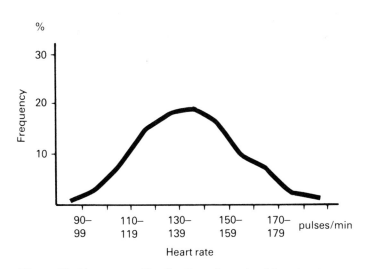

Figure 82. **Frequency distribution of maximal heart rates.**
Measured on 380 factory workers during the course of a working day in the German iron and steel industry, during the years 1961–1969. After Hettinger [113].

Hettinger [113] compared the energy expenditure (work kJ) with the work pulses of 552 workers in the German iron and steel industry. Related to maximum limits, values for work kJ were generally lower than the values for work pulses, a discrepancy which was attributed to the effects of static effort and heat. About one-third of the workplaces gave rise to work pulses above the limit of 40 beats/min.

Heavy work in agriculture

Despite mechanization, heavy work still exists in agriculture, and some studies by Brundke [34] may be mentioned in this connection.

A fruit grower, a farmer and a dairyman were studied, and their pulses taken throughout a working week. Resting pulse was measured when they were sleeping at night. Figure 83 shows the results of recording their work pulse.

It is clear from Figure 83 that the critical limit of 40 work pulses/min was exceeded on several occasions, especially by the dairyman and by the farmer at harvest time. These studies confirm that *in spite of mechanization, heavy work is still part of the programme for the agricultural worker.*

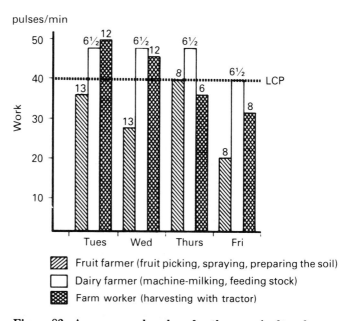

Figure 83. **Average work pulses for three agricultural occupations.** *The resting pulses were measured during night-time sleep. The figures at the top of each column indicate the approximate length of the working day in hours. LCP = limit of continuous performance for an 8-hour day. Drawn from data given by Brundke [34]*

Heavy work by women in the textile industry

There are many jobs in industry that involve heavy work without appearing to do so at first glance. One such example may be taken from the textile industry, where women have the

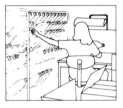  

Grasping motion   Checking and   Packing to the right 2 s
to the left 1–2 s   tying ends 10–12 s

Figure 84. Inspection of spools for artificial fibre.
The three drawings show the three operations involved. Weight of a
spool=3 kg. Performance: 730–960 × 3 kg=2200–2900 kg per arm
per shift. After Nemecek and Grandjean [199].

job of examining spools of artificial fibre and packing them
into boxes. The three key moves in this task are illustrated in
Figure 84.

Reach to the left varies between 500 and 900 mm while that
to the right amounts to about 900 mm. The movement across
from left to right requires an effort that is partly static
(supporting a weight of 3 kg), and partly dynamic, and each
turn is accompanied by slight bending and twisting of the
trunk. The factory introduced a '*new way*' of knotting the
ends of the fibres, and since this rationalization did not
produce the improved performance that was expected, the
question arose whether the operators might be physically
overstressed.

Because of this situation we selected five workers, and:

1. Took their pulse.
2. Counted how many spools they handled in 15 minutes.
3. Investigated the extent of bodily aches and subjective
   impressions of fatigue.

To assess the bodily aches and the fatigue, we used a
double-ended questionnaire, both before and after the task.

The resulting pulse measurements are shown in Figure 85.
Assuming that the desirable upper limit for continuous effort
for women is 30 work pulses/min, we can see that this has been
exceeded in 6 out of 10 cases, since we have twice recorded
work pulses of more than 40/min. *The cause is to be found in*
the increased demand for muscular effort from the arms.
According to Rohmert [217] the maximum lifting power of a
woman's outstretched arm averages 6 ± 1·5 kg, so that, to lift
a spool weighing 3 kg, she is almost certainly exerting more
than 50% of her maximum power. This is too high even for
dynamic effort, and obviously so for static effort, which ought
not to exceed 15–20% of maximum power. Moreover we have
not allowed for the weight of the outstretched arm itself,
which is well known to be between 19% and 30% of its
maximum muscular effort. So these studies have shown that

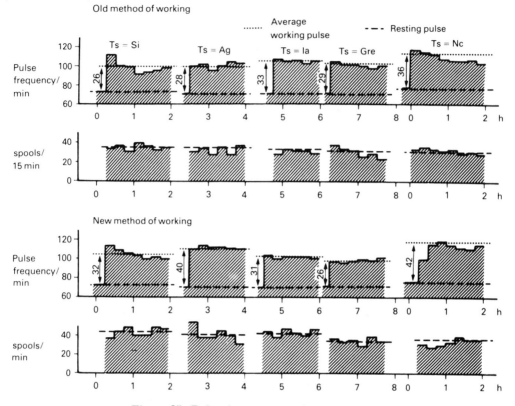

Figure 85. Pulse frequency and performance of five female operatives checking spools (bobbins) in a textile factory.
Ts = operative, designated by an abbreviation of her name. Broken line = resting pulse while sitting. Dotted line = average working pulse. Vertical arrows = work pulse averaged over 2 hours. After Nemecek and Grandjean [199].

this *apparently light form of work involves the operators in too much muscular effort.*

The assessment of bodily aches and fatigue disclosed that pains in the right arm, back and neck greatly increased during the working shifts. There is no doubt that these complaints are related to the postures that the workers had to adopt.

Finally, Table 16 summarizes the effect of the '*new method*', with its demand for increased performance, on the output, the work pulse and the level of aches and fatigue. The 'self-assessment' scheme made use of an average of the daily variations in replies to the questions involving fatigue, and aches in arms, hands, back and neck.

Although these results are not statistically significant, because of the small number of observations, the trend is obvious. The work load was already excessive under the '*old method*', and the '*new method*' — by reducing the inspection time, and involving more arm work to left and right—*has increased the work load still further, which tends to raise the*

Table 16. **Effect of the increased performance of the new method on output, work pulse, fatigue and aches and pains in five female textile workers.**
M = *method. After Nemecek and Grandjean* [199].

Worker	No. of spools handled per shift		Work pulses per minute		Increases in daily aches and fatigue (in mm of scale)	
	Old M	New M	Old M	New M	Old M	New M
Ne	730	959	36	42	1	18
Ag	720	912	28	41	22	27
Ge	732	840	29	26	7	19
Ja	732	960	33	31	3	31
Si	852	960	26	32	26	30

work pulse and apparently also to cause more fatigue and aches and pains in the operatives.

Conclusions

From the evidence of these results, the work could be lightened in the following ways:

1. Shortening the distance to be reached at each side.
2. Lowering the working level.
3. Introducing mechanical aids to reduce the load on the hands, e.g., a swivelling support for the spool.
4. Reorganizing the work, with a rotation between different operations.

7. *Handling loads*

7.1. Back troubles

Loss of work through back troubles

Lifting, handling and dragging loads involve a good deal of static effort, enough to be classified as heavy work. The main problem of these forms of work, however, is not the heavy loads on the muscles, but much more *the wear and tear on the intervertebral discs*, with the increased risk of back troubles. That is why the handling of loads deserves a chapter to itself.

Back troubles are painful and reduce one's mobility and vitality. They lead to long absences from work, and in modern times are among the main causes of early disability. They are comparatively common in the age group 20−40, with certain occupations (labourer, farmer, porter, nursing staff, etc.) being particularly vulnerable to disc troubles. Moreover, workers with physically active jobs suffer more from ailments of this nature, and their work is more affected than is the case with sedentary workers.

Over-exertion and lower back pain

Interesting statistics were published a few years ago by the US National Institute for Occupational Safety and Health (NIOSH) [202]. Over-exertion was claimed to be the cause of lower back pain by over 60% of the people suffering from it. Over-exertion injuries account for about one-quarter of all reported occupational injuries in the USA, with some industries reporting that more than half of the total reported injuries are due to over-exertion. Approximately two-thirds of over-exertion injury claims involved lifting loads, and about 20% involved pushing and pulling loads. An analysis of reports by HM Factory Inspectorate [44] by Pheasant [209] reveals that, between 1945 and 1980, 'handling goods' was cited as the primary cause of between 25 and 31% of all reported industrial injuries. Another UK report presents the following data: 61% of the over-exertion injuries (12.5% of all accidents) involve the back and 74% of these back injuries were due to lifting activities [260].

According to Krämer [146], disc troubles are in FR Germany the cause of 20% of absenteeism; and 50% of premature retirements.

All these figures show how widespread disc trouble is today, and justify the efforts of ergonomists to reduce the amount of wear and tear on the intervertebral discs by attention to posture. When the 'work seat' was discussed in Chapter 5 we considered the anatomy of invertebral discs and the nature of disc troubles. Here are a few supplementary remarks.

Disc troubles

It is well known that the vertebral column, or spine, has the shape of an elongated S: at chest level it has a slight backwards curve, called a kyphosis; and in the lumbar region it is slightly curved forwards, the lumbar lordosis. This construction gives the spine elasticity to absorb the shocks of running and jumping.

The loading on the vertebral column increases from above downwards, and is at its greatest in the lowest five lumbar vertebrae. As we have already said, each pair of vertebrae are separated by an intervertebral disc.

Degeneration of the discs first affects the margin of the disc, which is normally tough and fibrous. A tissue change is brought about by loss of water, so that the fibrous ring becomes brittle and fragile, and loses its strength. At first the degenerative changes merely make the disc flatter, with the risk of damage to the mechanics of the spine, or even of displacement of the vertebrae. Under these conditions quite small actions, such as lifting a weight, a slight stumble or similar incidents, may precipitate severe backache and lumbago.

When degeneration of the disc has progressed further, any sudden force upon it may squeeze the viscous internal fluid out through the ruptured outer ring, and so exert pressure either on the spinal cord itself or on the nerves running out from it. This is what happens in a case of a 'slipped disc' or disc herniation. Pressure on nerves, narrowing of the spaces between vertebrae, pulling and squeezing at adjoining tissues and ligaments of the joints are the causes of the variety of aches, muscular cramps and paralyses including lumbago and sciatica which commonly accompany disc degeneration.

Three different approaches can be distinguished among the studies evaluating the risk of back troubles when loads are lifted:

1. The measurement of intervertebral disc pressure.
2. Biomechanical models to predict compression forces on the lumbar spine.
3. The measurement of intra-abdominal pressure.

These three approaches will be summarized below.

7.2. Intervertebral disc pressure

Loading of the intervertebral discs

In Sweden, Nachemson [195, 196] and Andersson [5], have made a thorough study of the effects of bodily posture, and of

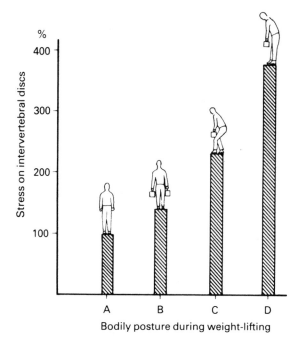

Figure 86. The effect of body posture when lifting weights on the intervertebral disc pressure between the third and fourth lumbar vertebrae.
A = upright stance. B = upright stance with 10 kg in each hand. C = lifting a load of 20 kg with knees bent and back straight (correct stance for weight-lifting). D = lifting 20 kg with knees straight and back bent. Pressure on discs during upright stance (A) is taken as 100%. After Nachemson and Elfström [196].

the handling of loads, on pressures inside the intervertebral discs.

Figure 86 shows the results of handling various weights on these pressures in nine people, two of whom had back troubles, and the other seven of whom were in good health. The figure shows clearly the effect of a bent back on the loading of the discs when a weight was being lifted.

Bending the back, while keeping the knees straight, puts a much greater stress on the discs in the lumbar region than keeping the back as straight as possible and bending the knees.

Lifting technique and disc pressure

Figure 87 shows how pressure develops in the discs during the two types of lifting action. This pressure curve shows very clearly how lifting a load with a bent back brings about a sudden and steep increase in internal pressure in the discs, and quickly overloads them, especially the fibrous rings.

Scientific studies confirm everyday experience that persons who have disc troubles are specially liable to sudden and violent pains, and even paralysis. These symptoms are pre-

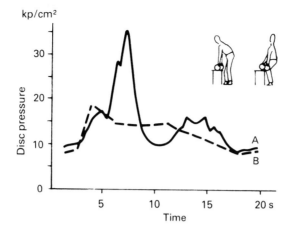

Figure 87. **Curves of pressure within the intervertebral disc between the third and fourth lumbar vertebrae, while lifting a load of 20 kg.** *A = back rounded, knees straight. B = back straight, knees bent. After Nachemson and Elfström* [196].

cipitated by sudden heavy loads placed on the discs, a risk which is increased by working methods involving unskilful manipulation.

If a person bends over until the upper part of his body is horizontal, then the leverage effect imposes very heavy pressure on the discs between the lumbar vertebrae. An average weight of the upper part of the body would be about 45 kg and the length of leverage about 350 mm, with a resulting moment of between 1000 and 2000 Nm. If a weight is lifted at the same time, the force on the discs could rise to 3000−4000 Nm.

The exact pressures per disc for various bodily postures are summarized in Table 17.

Distribution of loads on discs

When a rounded back causes curvature of the lumbar spine, the loads imposed on the intervertebral discs are not only

Table 17. **Loading of the disc between the third and fourth lumbar vertebrae in N, during various postures and tasks.** *After Nachemson and Elfstrom* [196].

Posture/activity	N
Standing upright	860
Walking slowly	920
Bending trunk sideways 20°	1140
Rotating trunk about 45°	1140
Bending trunk forwards 30°	1470
Bending trunk forwards 30°, supporting weight of 20 kg	2400
Standing upright holding 20 kg (10 kg in each hand)	1220
Lifting 20 kg with back straight and knees bent	2100
Lifting 20 kg with bent back and knees straight	3270

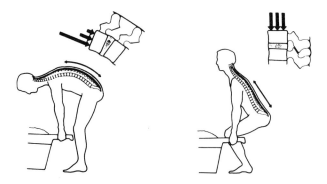

Figure 88. How the pressures on the intervertebral discs are distributed when a load is being lifted, with bent, and with straight back. *The round back (left) leads to heavy pressures on the front edge of the disc, and increases the risk of rupture. The straight back (right) ensures that the loads on the disc are evenly distributed, thus reducing wear and tear on the fibrous ring of the disc.*

heavy, but asymmetrical, being considerably heavier on the front edge than on the back (see Figure 88). *The resultant stresses on the fibrous ring are certainly bad for it, and must be regarded as an important factor in the 'wearing out' of the disc.* Further, it must be assumed that the viscous fluid inside the disc will tend to be squeezed towards the side with less pressure, in the back, with the danger that this fluid will leak towards the spinal cord. This is yet another argument for keeping the back as straight as possible when lifting a heavy load.

7.3. Biomechanical models of the lower back

Several authors [40, 186, 247] have developed biomechanical models to evaluate the compression forces and moments acting on the intervertebral discs of the spine. An important contribution was made by Chaffin [40], who recently, together with Andersson, published all his results in the book *Occupational Biomechanics* [42]. These authors used the hip moment to predict the expected abdominal pressure, the compression force as well as the shear forces on the spinal disc L5/S1 (fifth lumbar vertebra and os sacrum). The details of this procedure can be looked up in the book [42]; here we shall restrict ourselves to some of the results.

First of all it must be pointed out that the distance between the spine and the hands holding the load plays an important role, i.e., the lumbar spine is greatly affected if the load is moved closer to the torso or further away from it, as Figure 89 shows.

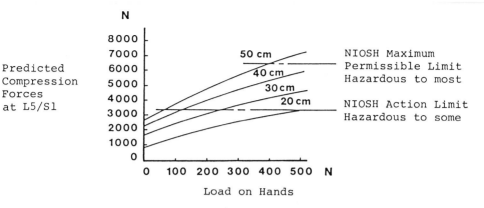

Figure 89. Predicted compression forces on L5/S1 of the lumbar spine for varying loads lifted in four different positions from body. *H = Distance from L5/S1 to the hand holding the load. According to Chaffin and Andersson* [42].

It is evident that the compression forces increase with the weight and distance of the load: when the lifted load is increased by 200 N the compression force increases by nearly 1400 N. Furthermore, the Figure reveals that an extension of the load distance from 300 to 400 mm is associated with an increase in the compression forces of about 1000 N.

Chaffin and Andersson [42] confirm the empirical rule which recommends lifting a load as close to the trunk as possible. On the other hand, they establish that a squatting posture in order to lift large loads from somewhere around the bent knees involves higher compression forces than lifting the same weight with a stooped lifting posture. In other words, if the load is too bulky to fit between the knees, stooping over the load might be more advantageous than squatting down to lift it. The biomechanical analysis showed that the shear forces, however, were larger in the stoop posture compared to the squat posture. This suggests that keeping the trunk in an erect posture is, whenever possible, to be preferred.

Recent studies have also dealt with the dynamic aspects of lifting loads, indicating that inertial forces increase the compression forces on the L5/S1 disc during the accelerative phase, and that static models underestimate spinal stresses. Mital and Kromodihardjo [186] developed a three-dimensional kinetic biomechanical model. The measurements include film records by two cameras, the calculation of reactive forces and moments at joint centres as well as spinal compressive and shear forces. This model was used to analyse compressive and shear forces under various task variables. The results revealed that lifting loads asymmetrically (i.e., involving a rotation of the trunk) or lifting large boxes or boxes without handles generated increased spinal stress.

7.5. Intra-abdominal pressure

Load lifting is always accompanied by a considerable increase in intra-abdominal pressure due to the contraction of the back extensor muscles (erector spinae) and the abdominal muscles. These forces stabilize the spine while lifting loads. A number of studies have shown a close correlation between the magnitudes of compression forces acting on the lower spine during load lifting and the magnitudes of intra-abdominal pressure rises.

The Davis and Stubbs studies

Davis and Stubbs [53, 54] measured the intra-abdominal pressure by using a capsule containing a pressure-sensitive element and a radio transmitter, which had to be swallowed by the subjects. It was concluded that during a load lifting effort, intra-abdominal pressure can be used to give an accurate indication of the spinal stress.

In an epidemiological study the authors observed that occupations in which peak intra-abdominal pressures of 100 mm Hg or more were induced, had an increased liability to reportable back injuries. They suggested a limit to the intra-abdominal pressure of 90 mm Hg, although values of 150 mm Hg are not uncommon in those who regularly lift weights.

Davis and Stubbs [53, 54] measured the intra-abdominal pressures with their pressure-sensitive capsules on a great number of subjects who had to lift different weights, assuming 36 different positions of the arms and different standing postures. One- and two- handed, frontal and sideward lifts were covered in repetitive and occasional activities.

The 90 mm Hg values were incorporated in contour maps [53], showing suggested limits for lifting forces for 95% of the population. For more details the reader is referred to the publication [53].

7.5. Recommendations

Acceptable loads for lifting

Maximum permissible weights to be lifted manually have been assessed by many authors and governmental agencies. 20 years ago the proposed maximum weights were to some extent deduced from maximum lifting power. They were higher than the limits proposed today, which primarily attempt to reduce the risk of back injuries to a minimum. This attempt raises several questions and problems. First of all, one must keep in mind that intervertebral disc injuries are in many cases an 'idiopathic disease', i.e., a degenerative process which is not caused by external factors. It is therefore obvious that maximum loads for lifting will hardly prevent the occurrence of intervertebral disc injuries. Furthermore, intervertebral discs will become less resistant to physical loads and especially sensitive to load lifting activities with increasing age. These considerations suggest that recommendations for acceptable loads as well as for adequate lifting techniques are nevertheless worthwhile.

Restrictions to load limits

Other relevant remarks and restrictions are pointed out by Pheasant [209]: "Industrial manual handling tasks are characteristically 'undesigned' activities, and have something of an extempore quality". For such occasional lifting tasks it will be difficult to set up standards. For repetitive and continuous load lifting the near future should provide help by mechanical solutions, such as robots or conveyor equipment. Finally it should be emphasized that many lifting tasks are associated with turning actions, which impose a rotation or twist upon the spine. Such lifting activities are particularly hazardous, and this has seldom been taken into consideration by current proposals for maximum permissible loads to be lifted.

The NIOSH guide

The US National Institute of Occupational Safety and Health (NIOSH) deduced limits for lifting activities from the studies by Chaffin [40]. These recommendations not only consider the horizontal distance of the load from the body but also the frequency of lifting, the vertical travel distance and height of the load at the beginning of lifting. Under optimum conditions 40 kg (392 N) can be lifted. NIOSH [202] has set a *maximum permissible limit* and an *action limit*. The former corresponds to the capacity of a 75th percentile man or a 99th percentile woman; the action limit corresponds to the 1st percentile of men or 25th percentile of women. These limits are indicated in Figure 89; the maximum permissible limit corresponds to a compression force at L5/S1 of 6400 N and the action limit is equivalent to 3400 N. The NIOSH Practices Guide [202] considers workloads exceeding the maximum permissible limit to be unacceptable and should be reduced; loads exceeding the action limit are considered a 'normal risk'; and loads between the two limits require selection and/or training of the workforce.

The NIOSH recommendations [202] are rather extensive and their details cannot be given here.

A major drawback of the NIOSH recommendations is that they only cover symmetrical two-handed lifting, performed directly in front of the body. In fact, most lifting activities on the shop-floor involve sideward movements, rotation of the trunk or some other asymmetrical elements.

Maximum acceptable forces by Davis and Stubbs

The results of the studies by Davis and Stubbs [53, 54] were adopted by the UK Ministry of Defence [185], which assessed standards for maximum permissible forces for lifting activities. These limits are based on the above-mentioned

Table 18. Maximum acceptable loads for young men while lifting, (N).
Frequency not more than once per min., and for a maximum intra-abdominal pressure of 90 mm Hg. For more frequent movements these values have to be reduced by 30%. The N values are rounded off by approximately 2%. After Davis and Stubbs [54].

Condition	Grasping distance, expressed as a fraction of arm's length			
	1/4	1/2	3/4	4/4
Standing up				
Two-handed lift, frontal	350	250	150	100
One-handed lift, frontal	300	220	140	100
One-handed lift, sideways	270	200	130	100
Seated				
Two-handed lift, frontal	270	170	120	110
One-handed lift, frontal	350	220	140	100
One-handed lift, sideways	330	210	140	90

maximum acceptable intra-abdominal pressure of 90 mm Hg. The acceptable limits cover a wide range of one- and two-handed exertions in standing, sitting and kneeling positions, occasional and frequent lifts for different groups and for both sexes.

A few examples of maximum acceptable forces for lifting loads are reported in Tables 18 and 19.

Table 19. Maximum acceptable loads under various lifting conditions, (N).
The loads given should be safe for 95% of the indicated age and sex groups. It is assumed that the activity will be performed in an upright standing position. Two-handed lifts should be carried out in front of the body; if not, loads should be reduced by 20% (according to Pheasant [209]). The N values are rounded off by approximately 2%. Occ. = Occasional lifts; less than once per min. Freq. = frequent lifts; more often than once per min. All tabulated data are taken from Davis and Stubbs [54].

Activity	Men				Women			
	Under 50		Over 50		Under 50		Over 50	
	Occ.	Freq.	Occ.	Freq.	Occ.	Freq.	Occ.	Freq.
Two-handed lift; compact load, close to the body, within preferred range of heights	300	210	240	140	180	130	140	100
One-handed lift; compact load close to the body	200	140	120	80	120	80	70	50

Some remarks by Pheasant [209] relating to Table 19 should be emphasized here: "the restriction of loads to within these (or any other) levels does not guarantee safety. ...There is no such thing as a safe load. An unfit person may injure his back (or more accurately trigger an attack of pain in his already degenerate spine) by reaching awkwardly to pick up the most trivial loads. A fit person may injure himself handling a very modest load if he slips and loses his footing. It is not possible to specify a load which guarantees safety". These considerations are in accordance with the reflections on the idiopathic character of intervertebral disc diseases (made at the beginning of this chapter), which modify the importance of maximum permissible load limits.

Practical hints

The following rules are based on general experience as well as on scientific knowledge:

1. Seize the load and lift it:
    with a straight back;
    with bent knees, as shown in Figures 88 and 90.
2. Get the load as close to the body as possible:
    by grasping the load whenever possible between the knees;
    by good foot placement, as shown in Figure 90.
3. Make sure that your hold on the load is not lower than knee height, the peak lifting forces being around 500 and 750 mm above the ground. A lift starting at knee height can be continued comfortably to a level of 900−1100 mm. Lifts starting at elbow height may be continued to shoulder height; higher levels require much more strength.
4. If the load does not have handles, one could tie a rope sling around the load and use a harness or hooks.
5. Loading ramps are suitable at levels around 500 mm, whereas an optimal zone for storage is at levels between 800 and 1100 mm.
6. Avoid a rotating or twisting movement when lifting a load.

Figure 90. Lift loads as close to the body as possible, with suitable foot placement.

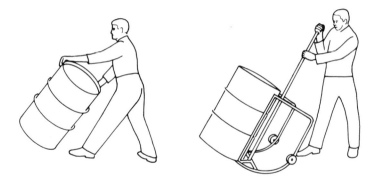

Figure 91. Handling casks.
Left: tilting and rolling, with the upper part of the body held upright.
Right: lightening the work by using a trolley.

7. Try, wherever possible, to use a trolley, a lifting ramp or similar mechanical aids. Such an example is shown in Figure 91.

Carrying techniques

When a heavy load is being carried, it is advisable to hold it vertically over the centre of gravity of the body, as far as possible. In this way the effort of balancing will be minimized, and unnecessary static muscular work avoided. The best way of carrying loads is by a yoke across the shoulders. It is less good to carry a burden in front of the body, where it places static loads on many muscles of the abdominal wall. Müller and Spitzer [191] have found that a workman can carry a load of 20 kg the following distances for an energy consumption of 1050 KJ:

using a yoke:	4·5 km
on his back:	3·9 km

The limits quoted in Table 19 should be used as guidelines for loads carried in one or two hands.

As we noted above in Chapter 6, a load of 50 kg is the most economical from the point of view of energetics, as well as being the most efficient when the return is taken into account. Hence 50 kg sacks are to be recommended over short distances. In contrast, 100 kg sacks are both bad physiologically and very much more harmful to the spine.

8. Skilled work

8.1. Acquiring skill

Skilled jobs call for a high degree of:

Quick and accurate regulation of muscular contraction.
Co-ordination of the movements of the individual muscles.
Precision of movements.
Concentration.
Visual control.

In practice, skilled work is mostly a matter for the hands and fingers only. The most important nervous processes that accompany a movement with an element of skill involved were shown in Figure 14. It is clear that to perform a delicate movement with speed and precision calls for a whole series of sensory nerve impulses, followed by motor directives from the brain.

Learning

During the period in which a skilled operation is being learned, we can distinguish between two processes:

1. *Learning the movements.*
2. *Adaptation of the organs involved.*

From a physiological point of view, *learning is essentially a matter of imprinting a pattern of the necessary movements upon the medulla of the brain*. To begin with, all the movements must be performed consciously, but as training progresses the conscious element is gradually reduced. New pathways and junctions are built up in the brain, and control of movements is gradually taken over completely by cerebral nerve centres. In simpler terms: *acquiring a skill consists mainly of creating reflex arcs which replace conscious control*.

   Learning to write is a good example. First the child learns one letter after another by consciously copying what the teacher draws. Writing is very laborious and difficult at this stage. Gradually, after months or years perhaps, the different letters, and later on complete words, become 'impressed on the brain as patterns for the necessary hand movements'. Writing has become 'automated', at least as far as finger movements

115

are concerned, and conscious awareness is diverted more and more to finding words and constructing sentences.

Another phenomenon appears during the learning phase: the gradual elimination of all muscular activity that is not essential to the skilled work in hand. The skilled man is relaxed and economical in his movements, whereas the novice's work is cramped and tiring. Hence the level of energy consumption for a given task decreases during the training period, as non-essential movements are gradually eliminated.

Recommenda-
tions for training

The following recommendations will make training easier:

1. *Short training sessions.* To acquire a skill calls for the highest concentration. A tired person easily develops bad habits, which are difficult to unlearn afterwards. It may be said that the higher the level of skill to be acquired, the shorter the training sessions should be. For very delicate skilled work four sessions per day, each of 15−30 minutes length, are enough to begin with, gradually becoming longer later on.

2. *Splitting the job into separate operations.* A work study should be made to enable the job to be divided into a number of distinct operations or processes. It is then possible to make tests to determine which parts of the job are the most difficult and to set the expected performance accordingly. It is also possible, and advantageous, to practise the most important operations by themselves, only later practising several operations in sequence, and especially switching from one to the other.

3. *Strict control and good examples.* As we have seen, the learner must avoid acquiring bad habits, so it is important that he or she should be strictly supervised throughout the training. Young people learn a good deal by direct and largely subconscious imitation, so they should be given highly skilled instructors to learn from. Older learners depend less on imitation and more on visual aids such as diagrams.

Adaptation of
organs

The second process in learning — adaptation of the bodily organs involved — is a matter of making gradual changes in the muscles and occasionally other organs, such as the heart or the skeleton. Muscular adaptation involves thickening the muscle fibres, and thereby increasing the total power of the muscle. Training for very rapid movement means not only increasing muscle power, but concurrently reducing internal friction by getting rid of some of the non-contractile material, such as connective tissue and fat.

8.2. Maximum control of skilled movements

With the object of increasing the precision and speed of skilled work, many authors since World War II have made detailed studies of several typical movements of the hand and forearm. The most important of their results include the following.

Position of the arms

Ellis [67] and Tichauer [256] studied how the height of the working surface affected manual performance. Ellis was able to confirm an old empirical rule: *the maximum speed of operation for manual jobs held in front of the body is achieved by holding the elbows down to the sides, and the arms bent at right angles.* In a similar situation in practice, Tichauer [256] studied 12 female packers of foodstuffs to find out how the position of the upper arms affected performance and working metabolism. The results, reproduced in Figure 92, showed that the best performance was achieved with the forearms held out sideways at an angle of 8−23° with the vertical.

In another study, Tichauer [257] found that, if the arms were held out as much as 45° to the sides, the shoulders took up a balancing posture, causing fatigue in the shoulder muscles. The need to do this was often a result of too low a seat.

Criteria often used to discover the optimum sequence of movements were the time consumed and the degree of precision; modern ergonomists would say that not enough attention was given to signs of fatigue and to monotony. With this reservation, the most important findings are summarized below.

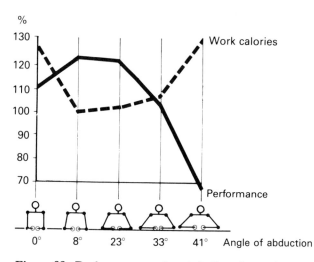

Figure 92. **Performance and metabolism for various attitudes of the upper arms (angles of abduction) when packing groceries.** *After Tichauer [256].*

Time and motion   Time studies show first of all that reaction time (interval
between the signal and the beginning of the motor reaction)
has a fairly constant value of 0·25 s, which is hardly affected
by either the nature of the movement or its distance from the
body. Hence the grasping time and grasping distance are not
quite linearly related, as Barnes [8] had already shown in
1936. Thus:

Grasping distance (mm)	Grasping time (%)
130	100
260	115
390	125

Other studies by Brown and his colleagues [32, 33] showed
that adjusting movements with the right hand were quicker
from right to left than from left to right. If the distance was
increased from 100 to 400 mm, the time for the movement
increased from 0·7 to 0·95 s.

Precision of   McCormick and Sanders [180] report some unpublished work
movement   by Briggs, showing that outward movements were more
precise than inward movements, and precision fell as the
distance of the movement increased. In similar fashion
Schmidtke and Stier [234] were able to show that the speed of
a horizontal movement with the right hand is greatest in a
direction 45° to the right, measured from a saggital plane in
front of the body, and slowest at 45° to the left. These studies
show that a hand operation can be performed faster and more
precisely if most of the movement comes from the forearm.
The movement is both slower and less precise if the upper arm
participates in it.

Optimal working   Experimental studies of the speed and precision of manual
field   operations are useful when deciding the layout of controls
in vehicles of all kinds, including aircraft, but they are less
significant when it comes to most industrial workplaces, since
the skilled operations involved are much more complex. The
work of Bouisset and his co-workers [23, 24] is more relevant
to this problem. These authors made their research subjects
shift weights of up to 1 kg, and studied the best conditions so
far as physical stress was concerned. Oxygen consumption
and electrical activity were measured in several muscles of the
arm and trunk, both for one-handed and two-handed actions.
The weight had to be moved over a distance of 300 mm, in
different directions, starting from a point immediately in front
of the centre of the body, and repeated 24 times/min. The
oxygen consumption for various directions is shown in Figure
93.
   *The lowest oxygen consumption corresponds to the direc-
tion of 60° from the front*, which agrees with the findings of
McCormick and Sanders [180] and Schmidtke [234]. This was

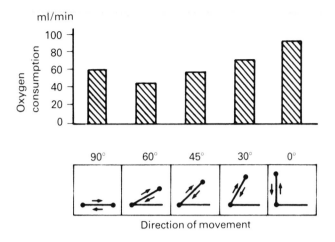

Figure 93. Energy expenditure and working field.
The test subjects must move weights of 1 kg repeatedly in the given directions. The direction of 60° is the most efficient, provided that the movement does not have to be controlled by eye. After Bouisset and Monod [23].

also the best angle for two-handed action, for all weights tested, and for all the different frequencies.

Electromyography explained these findings. Electrical activity in various muscles was distinctly less when the arms were held out to the side than when they were held forwards, the latter setting up heavy static loads in the arms, shoulders and trunk. These optimal angles cannot, however, be applied in practice without reserve. A control placed at an angle of 60° is ideal only when there is no need to look directly forwards at the same time. When forward vision is involved, or when both arms are used, angles of only a few degrees right or left are to be preferred.

Typing

Typing is skilled work, practised worldwide by millions of operators. Typing speed is usually very high, most operators manage up to five strokes per second. Typing, however, involves not only the control of the muscular activity in fingers, hands and arms, but also the sensory input, the integration and interpretation of the entering information and the transfer of the impulses to the muscular control centres where the typing movements for words are 'engraved' as complete movement patterns. That part of perception, interpretation and transfer of command can be considered the mental work in typing; it is certainly much more important than the automated process of operating the keys. It is well known that the time required for the mental part is considerably longer than the one for operating the keys. With modern typewriters typing performance is certainly no longer limited by typing speed but mainly by the mental load of the tasks. It is therefore useless to design keyboards with the aim to further accelerate the speed of typing. But there remain the postural problems of hands, wrists and forearms which could be solved with a proper keyboards design. These aspects have already been discussed in Chapter 5.

8.3. Design of tools and equipment used for skilled work

Handgrips

The design of hand grips has a high priority in skilled work. Hand grips which are not shaped to fit the hand properly, or which pay too little attention to the biomechanics of manual work, may lead to poor performance, and may even be injurious to the health of the operator. The hand and fingers are capable of a wide range of movements and grasping actions, which depend partly on being able to rotate the wrist and forearm. The most important of these many movements are illustrated in Figure 94.

The maximum grasping force can be quadrupled by changing over from holding with the fingertips to clasping with the whole hand. The power of the fingers is at its greatest when the hand is slightly bent upwards (dorsal flexion). In contrast, grasping power, and consequently level of skilled operation, is reduced if the hand is bent downwards, or turned to either side.

Tichauer [257] says that inclining the hand either outwards (ulnar direction) or inwards (radial direction) reduces rotational ability by 50%, and if the hands are held in such positions every day and frequently, this may cause inflammation of the tendon sheaths. This possibility was mentioned in Chapter 1, where Figure 9 shows an example of using a pair of wire-cutters at work. The conclusion from this is that, on

 Finger-tips   70 – 140 N

 Thumb and sides of fingers   70 – 140 N

 Clasping in thumb and fingers   300 – 540 N

Figure 94. Three ways of gripping with the hand.
The figures indicate the range of finger pressures, according to Taylor [252].

biomechanical grounds, *the hands should always be kept in line with the forearms as much as possible.*

*Design of
handgrips*

The work of Barnes [9] can be taken as a basis for the design of hand grips for the whole hand, with a cylindrical shape as the best. *This should be at least 100 mm long, and its effectiveness increases with thickness up to 30—40 mm.*

It is not possible to enumerate all the range of different tools that should be designed to suit the shape of the hand, but Figure 95 shows a few examples of good design.

Electric cutter

Hand shovel

Hand saw

Figure 95. Handgrips sensibly shaped to the anatomy and functioning of the hand.
Left: bad solutions
Right: good design.

*Hints and
equipments for
maximum skill*

Some of the important conditions for the maximum control of skilled movements have already been discussed earlier in this Chapter. The following 10 points summarize ways of making skilled work easier to perform:

1. The working field must be so arranged *that manual operations can be performed with the elbows lowered, and the forearms at an angle of 85—110°.*
2. *For very delicate work the working field must be raised up to suit the visual distance,* the elbows lowered, head and neck slightly bowed, and the forearm supported (in this context see the seven guidelines in Chapter 3).
3. *Skilled operations should not call for much force to be exerted,* since heavily loaded muscles are more difficult to control and to co-ordinate with others. Above all, avoid

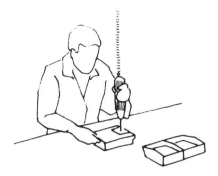

Figure 96. During skilled operations, the muscles should be kept free from having to exercise force, and especially from static effort.
The electric screwdriver is suspended from a spring support which reduces manual effort to a minimum.

imposing a static stress at the same time. It is equally bad to have to exercise skill immediately after physical effort. These recommendations are illustrated in Figures 96 and 97.

4. *Concentration on a manual operation is improved if it is not necessary to do other things at the same time.* Hence it is helpful to support the work, and if possible to have foot pedals with which to clamp and unclamp it, and to switch the machine on and off. Chutes are useful for taking finished work away and delivering fresh items.

5. The arrangement of working materials, parts and controls should be such as to *allow the operations to follow rhythmically in a sensible sequence.*

6. *A free rhythm is better than any kind of imposed tempo, whether it is time-controlled, or cyclical, or on a conveyor belt.* A free rhythm consumes less energy (because there are fewer secondary movements), motor control is easier, fatigue is reduced, and monotony and boredom less fre-

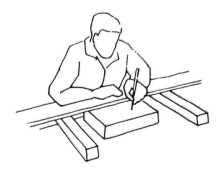

Figure 97. A good example of relieving muscular tension during skilled work.
Suitable supports for the forearms make the precision work of the lithographer much easier.

quent. It should be mentioned here that too slow a rhythm is bad because the work has to be supported, and too fast a rhythm is even worse, because of the nervous stress it imposes, and the fatigue it engenders. The operator usually finds his or her own rhythm instinctively, to suit his liking.

7. When both hands are used in the work, the working field should extend as little as possible to each side, to give the best visual control. The muscular effort should be symmetrical, i.e., as nearly as possible equal for the two hands, which should begin and end each movement together.

8. *Movements of the forearms and hands are at their most skilful, both in speed and precision, if they take place within an arc of 45−50° to each side.* Both for grasping things and for working, their optimum arc is two-thirds of their maximum reach, i.e., over a radius of 350−450 mm from the tip of the lowered elbow.

9. Horizontal movements are easier to control than vertical ones, and circular movements easier than zig-zag ones. Each operation should end in a good position for starting the next one.

10. *Handles of controls and tools should be shaped to fit the hand,* and should operate when the hand is held in line with the forearm.

9. Man–machine systems

A 'man–machine system' means that the man and the machine have a reciprocal relationship with each other. Figure 98 shows a simple model of such a system.

Obviously such a system is a closed cycle in which the man holds the key position because the decisions rest with him. The pathways of information and direction are, in principle, the following. The *recording display* gives information about the progress of production; the operator receives this information visually (*perception*), and must understand and assess it correctly (*interpretation*). On the strength of his interpretation, and in the light of his previous knowledge, he takes a *decision*. The next step is to communicate this decision to the machine by *using the controls*. A *control display* tells the operator the result of his action (e.g., how much water has been mixed in with the reagents). The machine then carries out the *production process* as programmed. The cycle is completed when various significant parts of the process, such as temperature or quantities, are displayed for the operator to see.

In a report for the World Health Organization, Singleton [243] points out that the machine is capable of high speed and precision, as well as being very powerful, whereas the man is sluggish, releasing only small amounts of energy; on the other hand, he is much more flexible and adaptable. Man and machine can combine to form a very productive system, provided that their respective qualities are sensibly used.

The control of machines was no great problem until recently, when the development of electronics, more elaborate controls and higher output, with the consequent need for accurate interpretation of the information displayed, made the operator's task both more delicate and more demanding. As a result, the 'human factor' in such systems became increasingly important. In an aircraft the speed of the pilot's or engineer's reaction can be vital; in chemical processes alertness and correct decision-taking may alone avert catastrophe. So modern man–machine systems need to be ergonomically sound.

The 'points of interchange' *from man to machine* and *from machine to man* — interfaces — are of paramount importance.

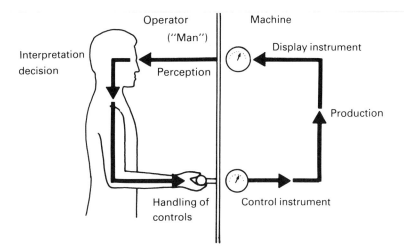

Figure 98. The 'man-machine system'.

The ergonomic interfaces of man—machine systems are therefore:

1. *Perception* of all the information on display.
2. *Manually operating the controls.*

Below we shall consider these problems from an ergonomic point of view.

9.1. Display equipment

A display conveys information to the human sense organs by some appropriate means. In man—machine systems it is usually a matter of visual presentation of dynamic processes, for example, fluctuations of temperature or pressure throughout a chemical production.

Display equipment used commercially falls into three categories:

1. *A window in which figures can be read directly.*
2. *A circular scale with moving pointer.*
3. *A fixed marker over a moving scale.*

Each of these has its advantages in certain circumstances, some of which are listed below.

Reading off values

If it is simply a matter of reading off the value of some quantity, the window display is best, provided that only one figure is visible at a time.

If it is necessary to see how a process is going on, or to note the amplitude of some change, then the pointer moving over a fixed scale gives more information. The same applies to a slow change, which needs to be checked from time to time. A

Type of display	 Moving pointer	 Fixed marker moving scale	Counter
Ease of reading	Acceptable	Acceptable	Very good
Detection of change	Very good	Acceptable	Poor
Setting to a reading: controlling a process	Very good	Acceptable	Acceptable

Figure 99. Different ways of displaying information.

moving scale with a fixed indicator mark may also be used for these purposes, but suffers from the disadvantage that it is not easy to memorize the previous reading, nor to assess the extent of movement.

If a process is to be set to some particular value (e.g., steam pressure or electrical voltage) it is easier to do this with a moving pointer, because a second, hand-operated pointer can be preset and the process accurately controlled as the two pointers come together. If the process goes slowly, then either of the other two types of instrument will serve equally well. Figure 99 illustrates the three types of dial and processes for which they are appropriate.

Fixed or moving scale

McCormick and Sanders [180] consider that in general a fixed scale with a moving pointer is better than the converse arrangement. The moving pointer catches the eye and is quick to read off. On the other hand fixed scales have their limitations. If they have to cover a wide range of values, the scale becomes too big and it is then better to see only a part of the scale moving against a fixed indicator.

Reading errors

In the years after World War II it was realized how important the layout of instruments in aircraft and transport vehicles is, in the interests of speed and accuracy of reading, and many studies were carried out to find the best design and arrangement. This usually involved the use of a tachistoscope, for which the criterion was the number of reading errors when the information was displayed for a fraction of a second.

Figure 100 shows results from a study made by Sleight [244]. Each of the scales was shown 17 times to a total of 60 research subjects, for a period of $0 \cdot 12$ s each. The results show significant differences in the frequency of reading errors, the 'open window' with $0 \cdot 5\%$ errors, being undoubtedly the best.

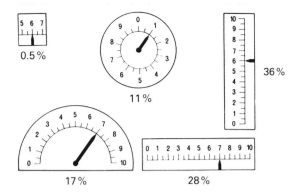

Figure 100. How the type of dial affects the precision of reading.
The figures give percentages of reading errors during a display time of 0·12 seconds. After Sleight [256].

It appears that with the open window no time is lost in locating the pointer.

Murrell [193] and other authors have cast doubt upon the suitability of the tachistoscope, because under practical conditions the operator looks at the indicator for as long as is necessary to read off the figure. In spite of these objections we can agree with McCormick and Sanders [180] that *in general the 'open window' type of instrument is less prone to reading errors than the others, and that a vertical scale is less satisfactory than a horizontal one.* McCormick and Sanders [180] do admit, however, that vertical scales are better when several need to be compared side by side.

Altimeters

An instructive example of the need for sound instrumentation is the design of altimeters, which have not infrequently been

Figure 101. An aircraft altimeter designed to reduce reading errors.
According to Hill and Chernikoff [117], *however, a weakness is that the pointer sometimes covers up one or other of the counters.*

the cause of aircraft accidents. The traditional altimeter had three pointers (hands), one reading hundreds of feet, the second thousands and the third tens of thousands. Many studies have shown that both in the laboratory and in flight, reading errors could be reduced by designing the altimeter differently. Hill and Chernikoff [117] were, in 1965, the first to arrive at the instrument which is currently the most trouble free and the most pleasant to use. It consists of a round dial with a single moving pointer for the 100-foot scale, and two windows, one of which shows the higher ranges and the other the setting for the atmospheric pressure (see Figure 101).

Matching instrument and information requirement

It is important that the instrument should give the operator only the information he requires, for instance by displaying the smallest unit that he is likely to read off. Thus if he needs to read pressures to the nearest 100 mm Hg, then the smallest division should be 100 mm Hg.

Sometimes the instrument is required to show not a precise reading, but a range, say between a lower, safe limit and an upper danger line, or 'cold', 'warm' and 'too hot'. Here a moving pointer is best, and the various ranges should be picked out in different colours, or patterns. Figure 102 shows an example of an instrument on which information is kept to a minimum.

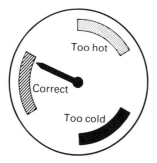

Figure 102. A display instrument should convey the required information as simply and unmistakeably as possible.

Scale graduation

Even more important than the shape of the dial itself is the size of the scale graduations. Since lighting and contrast are not always ideal, and since other adverse factors are present at a real work place, we recommend rather big graduations as follows: if a is the greatest viewing distance to be expected, in mm, then the minimum dimensions of graduations ought to be as follows:

Height of biggest graduations:	$a/90$
Height of middle graduations:	$a/125$

Height of smallest graduations:	$a/200$
Thickness of graduations:	$a/5000$
Distance between two small graduations:	$a/600$
Distance between two big graduations:	$a/50$

Recommenda-
tions for the
design of scale
graduations

From the above discussion, plus one or two other obvious considerations, the ergonomic design of scale graduations can be summarized as follows:

1. Height, thickness and distance apart of scale graduations must be such that they can be read off with minimum likelihood of error, even if lighting conditions are not ideal.
2. The information presented should be what is actually wanted: scale divisions should not be smaller than the accuracy required; qualitative information should be simple and unmistakable.
3. Scale graduations should give information that is easy to interpret and to make use of. It is laborious to have to multiply the reading of the instrument by some factor, and if this is unavoidable, then the factor should be as simple as possible, say × 10 or × 100.
4. Subdivisions should be 1/2 or 1/5: anything else is difficult to read off.
5. Numbers should be confined to major scale graduations and, once again, subdivisions should be 1/2 or 1/5.
6. The tip of the pointer should not obscure either the numbers or the graduations, and if possible should not be broader than a scale line. It is best if the tip of the pointer comes as close as possible to the scale, without actually touching it. Figure 103 shows good and bad pointers.
7. The pointer should be as nearly as possible in the same plane as the graduated scale, to avoid errors of parallax, and the eye must be positioned so that the line of sight is at right angles to the dial and pointer.

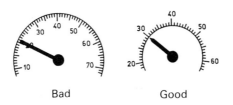

Bad   Good

Figure 103. Bad and good arrangements of numbers and pointer on a dial.
The tip of the pointer should be only as broad as one of the scale lines, and it should not obscure the number.

Letters and
figures

Western literature contains many studies of the effects of size and shape of letters and figures on visual perception, but these have mostly concerned aircraft. A good review of this literature will be found in the books by McCormick and Sanders [180] and Murrell [193]. The current view can be summarized as follows:

The question whether white letters on a black background are or are not easier to read than the opposite is still not fully decided.

Black letters on a white background are to be preferred, in principle, because white letters tend to blur, and a black background may set up relative glare against its lighter surroundings. (This is true both for printed and electronically produced text.) On the other hand white symbols show up better in poor lighting, especially if they and the pointer are luminous.

Size of symbols Sizes of letters and figures, thickness of lines, and their distance apart must all be related to the viewing distance between the eye and the display. The following formula may be used:

$$\text{height of letters or figures in mm} = \frac{\text{viewing distance in mm}}{200}$$

Table 20 gives examples.

Table 20. Recommended heights of lettering.

Distance from eye (mm)	Height of small letters or figures (mm)
Up to 500	2·5
501—900	5·0
901—1800	9·0
1801—3600	18·0
3601—6000	30·0

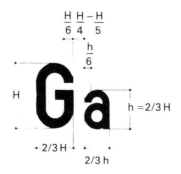

Figure 104. Recommended proportions for letters or figures.
H = height of capitals: h = height of lower case letters. The absolute sizes recommended are listed in Table 20.

Capitals and lower case are easier to read than letters all of the same size. Most letters and symbols should have the following proportions:

Breadth:	2/3 of height
Thickness of line:	1/6 of height
Distance apart of letters:	1/5 of height
Distance between words and figures:	2/3 of height

These proportions are illustrated in Figure 104.

The recommended dimensions of characters on VDTs will be discussed later in Chapter 17.

9.2. Controls

Adequate controls

Controls constitute a second 'interface' between man and machine. We may distinguish between:

1. *Controls which require little manual effort*: push-buttons, tumbler switches, small hand-levers, rotating and bar knobs; all of which can easily be operated with the fingers.
2. *Controls which require muscular effort*: hand-wheels, cranks, heavy levers and pedals; which involve the major muscle groups of arms or legs.

The right choice and arrangement of controls is essential if machines and equipment are to be operated correctly.
The following guidelines should be followed:

1. Controls should take account of the anatomy and functioning of the limbs. *Fingers and hands should be used for quick, precise movements; arms and feet for operations requiring force.*
2. Hand-operated controls should be easily reached and grasped, between elbow and shoulder height, and in full view.
3. Distance between controls must take account of human anatomy. *Two knobs or switches operated by the fingers should not be less than 15 mm apart; controls operated by the whole hand need to be 50 mm apart.*
4. Push-buttons, tumbler switches and rotating knobs are suitable for operations needing little movement or muscular effort, small travel and high precision, and for either continuous or stepped operation (click-stops).
5. Long-armed levers, cranks, hand-wheels and pedals are suitable for operations requiring muscular effort over a long travel, and comparatively little precision.

There is an abundant literature dealing with the ergonomic design and layout of controls. Good reviews are given by McCormick and Sanders [180], Kroemer [148], Morgan *et al.* [188], Schmidtke [233], Woodson and Conover [277] and in

DIN 33 401 [58]. The practical recommendations from these sources are summarized as follows.

Coding

Big machines in industry, agriculture and transport often have control panels with many similar controls and it is important to be able to pull the right lever or turn the right knob, even without looking at it. According to McFarland [183] the American airforce in World War II suffered 400 crashes in 22 months because the pilots mistook some other lever for that controlling the undercarriage.

For this reason *any controls that might be mistaken for each other should be so designed that they can be identified without difficulty.* Identification can be assured by:

1. *Arrangement.* For example, sequence of operation, or the difference between vertical and horizontal movement. Only a small number of controls can be identified in this way.
2. *Structure and material.* Figure 105 shows knobs of 11 different shapes, developed from experiments by Jenkins [131]. These were the shapes that were least often confused by blindfolded operators. Besides their shape and size, knobs can be made still more distinctive by their surface texture (smooth, ridged, etc.). These characteristics are most helpful if the control must be handled unseen, either in darkness, or while attention is being directed elsewhere;
3. *Colour and labelling* can be useful, but only in good light and under visual control.

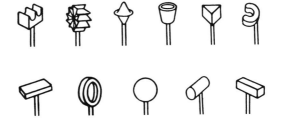

Figure 105. Types of hand grips (knobs) that are easy to distinguish. *After Jenkins [131], modified from Kroemer [148].*

Distance apart

If controls are to be operated freely and correctly, without unintentionally moving adjoining controls, they must be a certain minimum distance apart. Table 21 shows minimum and optimum separations.

Resistance

Controls should offer a certain amount of resistance to operation, so that their action is more positive and they are less likely to be triggered off by a slight tremor of the finger.

Table 21. Distance apart of adjoining controls.

Control	Method of operation	Distance apart (mm) Minimum	Optimum
Push button	With one finger	20	50
Toggle switch	With one finger	25	50
Main switch	With one hand	50	100
	With both hands	75	125
Hand wheel	With both hands	75	125
Rotating knob (round or bar-shaped)	With one hand	25	50
Pedal	Two pedals with same foot	50	100

Schmidtke [233] recommends the following resistances:

Single-handed rotation:	$2 \cdot 0 - 2 \cdot 2$ Nm
Pressing with one hand:	$10 \cdot 0 - 15 \cdot 0$ Nm
Pedal pressure:	$40 \cdot 0 - 80 \cdot 0$ Nm

Higher resistances than these are sometimes desirable to isolate particular controls from the others.

Among *controls suitable for precision operation with little force*, are the following: tumbler switches, hand levers and knobs.

Push-buttons for finger or hand operation

Push-buttons to be operated by the pressure of either finger or hand take up little space and can be made distinctive by colour or other marking. The surface area of the knob must be big enough for the finger or hand to be able to press the knob easily and apply the necessary pressure without slipping off. Recommended dimensions are:

Diameter:	$12 - 15$ mm
For an isolated emergency stop:	$30 - 40$ mm
Travel:	$3 - 10$ mm
Resistance to operation:	$2 \cdot 5 - 5 \cdot 0$ N

Finger-operated push-buttons should be slightly concave, whereas buttons to be pushed by hand should be mushroom-shaped. Dimensions of the latter should be:

Diameter:	60 mm
Travel:	10 mm
Resistance to operation:	10 N

Toggle switches

Toggle switches are easily seen and reliable in operation. They should preferably have only two positions ('off' and 'on'). Several toggle switches can be placed side by side, provided that each is clearly identified.

The direction of travel should be vertical, and the 'off' and

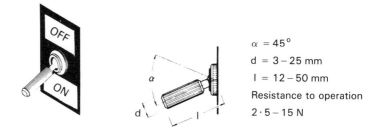

$\alpha = 45°$

d = 3 − 25 mm

l = 12 − 50 mm

Resistance to operation

2·5 − 15 N

Figure 106. Dimensions of toggle switches.

'on' positions marked above and below (convention whether the 'on' position is up or down varies in different countries). If a toggle switch with three positions is used, there should be at least 40° of travel between two adjoining positions, which should all be clearly identified. Recommended dimensions for toggle switches can be seen from Figure 106.

Hand levers

When the toggle is longer than 50 mm, it is spoken of as a *hand lever* and allows greater force to be exerted than a mere toggle. Here, too, the direction of movement should always be either up−down or forwards−backwards. When a hand lever has several positions, not just 'on' and 'off', each position should have a definite notch. If the lever is capable of fine adjustment, then some suitable form of support should be given to the elbow, forearm or wrist. Hand levers may have different knobs, according to their function:

A finger grip, diameter 20 mm.
A grip for the palm, diameter 30−40 mm.
A mushroom-shaped grip, diameter 50 mm.

Figure 107 shows a hand lever suitable for delicate control movements with the fingers. Hand levers which need considerable force to operate them are called *switch levers*, and come into the category of heavy controls. Suitable dimensions for a switch lever are indicated in Figure 108.

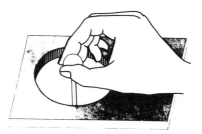

Figure 107. Control knob for precise operation with the fingers.
Lower arm and wrist should be supported by a smooth surface, on which they can easily move.

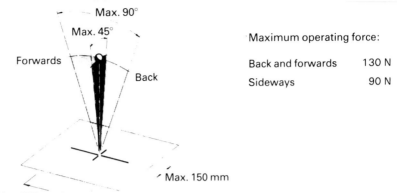

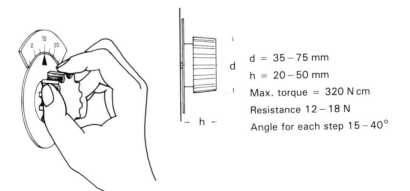

Figure 108. Large switch lever requiring moderate forces to operate it, with ranges of movement.

Rotating knobs

Rotating knobs may take a variety of shapes: round, arrow-shaped, combinations of knob and handle, or even several knobs on the same spindle.

An important requirement for all types is that *they must fit the hand comfortably and turn easily, and they should be in full view throughout the operation.*

Rotating switches

Figure 109 summarizes data about rotating switches with click stops. These should have a somewhat higher resistance than knobs for continuous rotation, so that the operator receives a clear 'tactile signal' at each position. A resistance level of $0 \cdot 15$ Nm is recommended. Furthermore, the successive positions should be not less than $15°$ apart if the switch is kept in view, and at least $30°$ apart if it is operated solely by feel.

Figure 109. Knob for step-by-step adjustment (click stops).
The notched margin makes it easier to control.

Knobs for continuous rotation

Knobs without click stops are suitable for fine and precise regulation over a wide range. An arc of 120° can be turned without shifting the grip, and under precise control; to turn further, the grip of the hand can be changed without difficulty. Such a knob can be turned either with the fingers or with the whole hand, and it helps if the surface of the knob is slightly grooved or roughened. The following dimensions can be recommended:

Diameter for use with two or three fingers: 10−30 mm
Diameter for use with the whole hand: 35−75 mm
Depth for finger control: 15−25 mm
Depth for hand control: 30−50 mm
Maximum torque for small knobs: 0·8 Nm
Maximum torque for large knobs: 3·2 Nm

Pointed bar knobs

Arrow-shaped or pointed bar knobs have the advantage of locating the reading quickly and easily, and should be 25−30 mm, measured along the bar. Figure 110 shows such a bar for use with click-stops.

Controls which call for muscular strength and long travel, but not a high degree of precision, can be operated by a crank, handwheel or pedal.

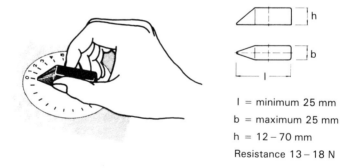

l = minimum 25 mm
b = maximum 25 mm
h = 12 − 70 mm
Resistance 13 − 18 N

Figure 110. Pointed bar knob for click stops.

Cranks

Geared levers, or cranks, are appropriate for either setting a control, or providing continuous adjustment, when a wide range of movement (long travel) is necessary. The gearing may be coarse or fine, according to the degree of precision required. The crank can be operated more quickly if the handle can rotate on its own axis, but a fixed handle is more precise. The following dimensions are recommended:

Length of lever arm for low torque
(up to 200 rpm): 60−120 mm
Length of lever arm for high torque
(up to 160 rpm): 150−220 mm
Length of lever arm for quick setting: up to 120 mm

According to Morgan *et al.* [188] the following are the rotation speeds (rpm) in relation to the crank radius:

mm	rpm
20	270
50	255
120	185
240	140

Hence it appears that the slower the speed of rotation of the crank, the longer it should be, and vice versa.

The relation between crank length and resistance to operation should be:

Crank 240 mm long, resistance $0 \cdot 5 - 2 \cdot 5$ Nm
Crank 120 mm long, resistance $3 \cdot 0 - 4 \cdot 0$ Nm

This range, 120—240 mm, is the best crank length for fine and precise control.

Dimensions of hand grips should be as follows:

Diameter:	25—30 mm
Length for one-handed operation:	80—120 mm
Length for two-handed operation:	190—250 mm

Handwheels

Handwheels are recommended when large forces must be applied, because they allow the use of both hands and a relatively long leverage where the turning speed is low.

Notches inside the rim of the wheel give a surer grip and allow force to be applied more efficiently.

Pedals

Pedal controls do not often have pedal pressures of more than 100 N, although some such occur in machinery used for agriculture, building and occasionally in industry. The brake pedals of motor vehicles also fall into this category. Pedals are very suitable for this purpose because the human foot is capable of very high pressures, up to 2000 N.

In order to exert heavy pedal pressures, it is necessary to have:

A high back rest.
Angle at knee between $140°$ and $160°$.
Angle at ankle $90-100°$.
Slope of lower leg $20-30°$

For even higher pedal pressures the whole leg must be brought into use ('leg pedal' rather than foot pedal), and the initial resistance must be enough to support the weight of the leg resting on the pedal.

Pressure should be applied with the ball of the foot, and the axis of tread should be a line from the ankle to the point of pressure on the backrest.

The following recommendations are made for *pedals with a heavy tread*:

Travel of pedal: 50—150 mm
(the smaller the knee angle, the longer the travel)
Minimum resistance to operation: 60 N

Pedals, such as the accelerator of a car, which require only light pressures, are operated solely with the foot, and their initial resistance is low. It is a help if this type of pedal can be operated with the heel on the floor and the foot resting on the lower edge of the pedal [149]. The following are recommended dimensions:

Travel of pedal: at most 60 mm
Maximum pedal angle: 30°†
Optimum pedal angle: 15°
Resistance to operation: 30—50 N

Pedals of all types should have a non-slip surface.

As already mentioned in Table 21, the distance apart (clearance) of adjacent pedals should be between 50 and 100 mm. Under special circumstances, e.g., for use with heavy shoes or gumboots, the pedals must be even wider apart.

Pedals for standing use

When a machine has a multiplicity of controls, foot pedals are sometimes used to relieve the hands, but this is undesirable if the operator stands at his work. In any case they should be restricted to operating an on-and-off switch. If a pedal cannot be avoided, then it should be of the type recommended in Figure 111.

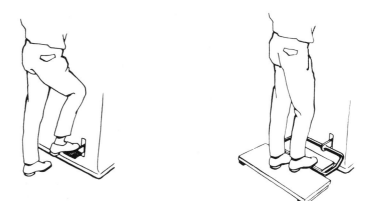

Figure 111. **Pedals are undesirable for standing work, since they set up heavy static loads in the legs.**
Left: bad arrangement, heavy loads on one leg.
Right: a good arrangement, a treadle that can be operated with either foot, at will.

---

† Pedal angle = operating angle between the two extreme positions.

9.3. Relationship between controls and display instruments

Relative speeds of movement

To set up a particular reading on a display instrument, the operator starts by moving the control quickly, until the reading is approximately correct, then he moves the control slowly to make the precise adjustment. The relative distances of travel of the control lever or knob and of the pointer of the instrument are of importance during this second, precision stage.

A coarse adjustment is easier when the pointer moves more quickly than the control, but *for precise adjustment the control should travel faster than the pointer*. The best ratio varies so much in different situations that it is not possible to formulate quantitative rules that would apply generally. Here we shall merely quote a recommendation of Shackel [238] for precise adjustment using a rotating knob and a pointer moving over a scale. One complete rotation of the knob should make the pointer move through 50—100 mm, allowing a reading tolerance of $0 \cdot 2$—$0 \cdot 4$ mm. For a higher reading tolerance of $0 \cdot 4$—$2 \cdot 5$ mm the pointer movement should be 100—150 mm per round.

Stereotyped reactions

When steering an unfamiliar motor car, we have a right to expect that turning the steering wheel clockwise will turn the wheels to the right: no-one would expect to turn the wheel to the left in order to steer to the right. Similarly, it is reasonable to expect that when a control knob is turned to the right, the pointer of the corresponding instrument will also move to the right. Such expectations depend on so-called *stereotyped reactions*, or fixations: experience has engraved the corresponding pattern on the brain, as we have already indicated during the discussion on learning to write (Chapter 8). *Stereotypes are thus conditioned reflexes which have become subconscious and 'automatic'.*

National differences

It must be pointed out here that some stereotyped reactions might be influenced by cultural traditions. That is why some arrangements of equipment must be adapted to national conventions, e.g., to national stereotyped reactions. Among many examples are electrical switches, which should always be 'on' in the same direction, but this is 'up' in some countries and 'down' in others.

Furthermore, not all stereotypes are equally firmly established and occasionally there are considerable individual deviations. Thus, layouts designed for right-handed operation are not equally suitable for left-handed people. Finally, another important exception to the rule is that clockwise rotation means an increase in whatever is being controlled. But both water and gas are controlled by stopcocks which turn off clockwise. This can lead to problems when one person has to control water/gas and electricity at the same time. In such a case some solution must be sought that involves the least risk of accidents and some degree of conscious control.

Controls and corresponding displays

Below we shall briefly discuss some rules which are valuable for industrial equipment all over the world. An important rule recommends that controls and instruments which are functionally linked ought to make corresponding movements that seem sensible to our own stereotypes. Figure 112 shows examples of sensible co-ordination between the direction of movement of controls and instruments with vertical or horizontal scales.

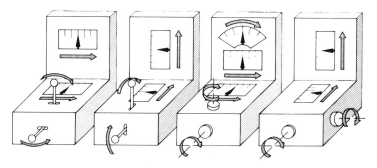

Figure 112. Examples of logical relationship between movements of controls and movements of indicators.

Some rules can be summarized as follows:

1. When a control is moved or turned to the right, the pointer must also move right over a round or a horizontal scale: on a vertical scale the pointer must move upwards.
2. When a control is moved upwards or forwards, the pointer must move either upwards or to the right.
3. A right-handed or clockwise rotation instinctively suggests an increase, so the display instrument should also record an increase.
4. Hoyos *et al.* [121] recommend that a moving scale with a fixed indicator should move to the right when the control is moved to the right, *but* the scale values should increase from *right to left*, so that a rotation of scale to the right gives increased readings.
5. When a hand lever is moved upwards, or forwards, or to the right, the display readings should increase, or a switch should move to the 'on' position. To reduce the reading, or to switch to 'off', it is instinctive to pull the lever towards the body or move it to the left, or downwards. These stereotypes affecting a hand lever are illustrated in Figure 113.

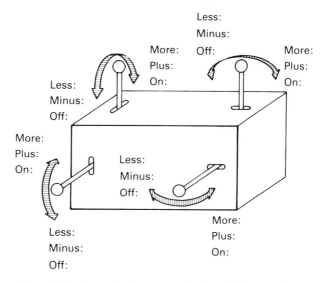

Figure 113. Expected movements in relation to the movements of a hand lever.
After Neumann and Timpe [200].

Sensible control panels

Problems of stereotyped reactions might also come up when designing large control panels. A sensible layout of both controls and display instruments will make supervision easier and reduce the risk of confusion caused by false readings.

Five principles should be taken into account when designing control panels:

1. As far as possible, display instruments should be located close to the controls which affect them. The controls should be placed either below the display, or, if need be, to the right of it.
2. If it is necessary for controls to be in one panel and display instruments in another, then the two sets should be set out in the same order and arrangement.
3. Identification labels should be placed *above* the control, and identical labels *above* the corresponding display.
4. If a number of controls are normally operated in sequence, then they and the corresponding displays *should be arranged on the panel in that order, from left to right.*
5. If, on the other hand, the controls on a particular panel are not operated in a regular sequence, then both *they and the corresponding displays should be arranged in functional groups*, in order to give the panel as a whole some degree of orderliness. The grouping can be emphasized by choice of colours, by labelling, by using knobs of different sizes and shapes, or simply by arranging members of a group in a row. Those controls and displays that are used most often should be directly in front of the operator, the less important ones to each side.

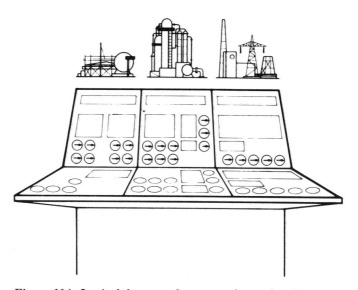

Figure 114. Logical layout of a control panel, with controls and display instruments in associated groups.
After Neumann and Timpe [200].

Figure 114 shows an example of a logical grouping of controls and displays on a control panel.

These recommendations may seem trivial and pedantic to the reader, but we must remember that long and often monotonous work can cause fatigue, and lead to reduced alertness and an increase in errors. Under these conditions a logical layout of controls and displays, which takes advantage of stereotyped, 'automated' behaviour, is beneficial, since a weary operator tends to fall back upon his conditioned reflexes.

10. Mental activity

10.1. Elements of 'brain work'

What constitutes mental activity?

Until a few years ago there was a simple line of demarcation between manual work, performed by operatives, and brain work which was the domain of white-collar workers; but nowadays this distinction is less clear on two grounds. Some jobs call for a good deal of mental activity without really coming into the category of brain work, e.g., information processing, supervisory work, taking important decisions on one's own responsibility. Moreover, this sort of work is by no means restricted to white-collar workers but is often delegated to manual operatives.

Hence the expression 'mental activity' as a general term for any job where the incoming information needs to be processed in some way by the brain. Such activity can be divided into two categories:

1. Brain work in the narrow sense.
2. Information processing as part of the man—machine system.

Brain work

Brain work in the narrow sense is essentially a thought process which calls for creativity to a greater or lesser extent. As a rule the information received must be combined with knowledge already stored in the brain and committed to memory in a new form. Decisive factors include knowledge, experience, mental agility and the ability to think up and formulate new ideas. Examples include: constructing machines, planning production, studying the files and extracting the essential facts from them, making a précis of these, giving instruction and writing reports.

Information processing

Information processing as part of man—machine systems has been discussed in Chapter 9, but nevertheless the essentials will be summarized again. They are:

$$\left.\begin{array}{l} \text{Perception} \\ \text{Interpretation} \\ \text{Mental processing} \end{array}\right\} \begin{array}{l} \text{of the information transmitted by} \\ \text{the sense organs} \end{array}$$

This 'processing' consists of combining the new information with what is already known, so providing a basis for decision-taking.

The mental load at work places such as we have discussed is conditioned by the following:

1. *The obligation to maintain a high level of alertness over long periods.*
2. *The need to take decisions* which involve heavy responsibility for the quality of the product and for the safety of work people and plant.
3. Occasional lowering of concentration *by monotony*.
4. *Lack of human contacts*, since one workplace is often isolated from any others.

The barrier that confronts us

Neurophysiology, psychology and other branches of science try very hard to get some degree of insight into the basic processes of mental effort. The well-known neurophysiologist Penfield makes the following comparison: 'Anyone who studies mental processes is like a person who stands at the foot of a mountain range. He has cut himself a clearing on the lowest foothill, and from there he looks towards the mountain-top, since that is his objective, but the summit is obscured in dense cloud'. It is this 'clearing' that we are about to examine.

Mental performances that are important in ergonomics include:

Uptake of information.
Memory.
Sustained alertness.

10.2. Uptake of information

Information theory

The information theory of Shannon and Weaver [240] has made an important contribution to the understanding of how information is taken up. These authors made a mathematical model that would represent quantitatively the transfer of information, and they devised the term '*bit*' (binary unit) for an item of information. The simplest definition of a bit is that it is the quantity of information conveyed by one of two alternative statements. For example: in olden times one flash of light from a watchtower might mean 'enemy approaching from the sea', whereas two flashes would mean 'enemy approaching from the land'. These alternative pieces of information are a bit.

As soon as there are more than two choices, of varying probability, the situation becomes much more complicated. So this theory has its limitations when applied to human beings, since the full significance of a stimulus conveying information cannot be interpreted by information theory. This theory is valid only for comparatively simple situations which can be split into units of information and coded signals. It is already useless when applied, for example, to the information being received by the driver of a car.

We need not go further into information theory at this point, since the interested reader needs only to consult Abramson [2].

Channel capacity theory

Another theory is based upon comparison of information uptake with the capacity of a 'channel' (channel capacity theory). According to this theory the sense organs deliver a certain quantity of information to the input end of the channel, and what comes out at the other end depends upon the capacity of the channel.

If the input is small, very little of it is absorbed by capacity of the channel, but if the input rises, it soon reaches a threshold value, beyond which the output from the channel is no longer a linear function of the input. This threshold is called the 'channel capacity' and can be determined experimentally for a variety of different sorts of visual and acoustic information.

Human beings have a large channel capacity for information communicated to them by the spoken word. Thus it has been calculated that a vocabulary of 2500 words requires a channel capacity of 34−42 bits per second [210]. This capacity is very modest when compared with the channel capacity of a telephone cable, which can handle up to 50 000 bits per second.

In everyday life the incoming information is much greater than the 'channel capacity' of the central nervous system, so that a considerable 'reduction process' must be carried out. It is estimated that this results in the following number of bits being characteristic of different parts of the system.

Process	Information stream in bits/s
Registration in sense organ	1 000 000 000
At nerve junctions	3 000 000
Conscious awareness	16
Lasting impression	0·7

Although these are only rough figures [248], it is clear that only a minute fraction of the information available is consciously absorbed and processed by the brain. The brain selects this minute fraction by reduction, i.e., by some kind of filtration process, about which we know virtually nothing.

10.3. Memory

Memory is the process of storing incoming information in the brain, often of only a selected portion of it, after it has been processed. How this selection is carried out, we do not know. But we do know that the process is subject to the emotions of the moment, and we must further assume that information to be stored must have some relevance to what is already there. Each person determines to a great extent what is relevant to himself, and what is not.

Two kinds of memory can be distinguished:

1. *Short-term or recent memory.*
2. *Long-term memory.*

Short-term memory comprises immediate recollection of instantaneous happenings, up to the remembrance of events which occurred a few minutes, or an hour or two ago.

Recall of events months or years after they occurred becomes long-term memory.

Items of information which become part of the memory leave 'traces', or *engrams* in certain areas of the brain. Information thus stored can be recalled at will, although, regrettably, not always as fully as one could wish!

There is a model for short-term memory. The information received leaves a 'trace', which continues to circulate as a stimulus inside a network of neurones, and which, by a kind of feedback, can be recalled into the conscious sphere at any time within a few hours.

This reservoir of short-term memory is expendable. This is what happens during 'retrograde amnesia', following some mechanical or emotional disturbance to the brain, whereby the memory of events immediately before the shock is obliterated. Periods of hours or even weeks can be lost from memory in this way. We must assume, therefore, that there is a period during which memories are being consolidated, or 'engraved' in the brain, and that during this period they are vulnerable to destruction. Afterwards they become more stable and surprisingly resistant engrams, which consitute long-term memory.

*The role of the
limbic system in
memory*

Experiments on animals, as well as clinical observations on humans, have shown that parts of the *limbic system* of the brain, especially that in the hippocampus, play an important role in the engraving process of short-term memory.

Before we go into that, let us take a general look at the anatomy of the brain, which may be considered in three sections:

1. The *fore brain*, including the cerebral cortex, where, among

other functions, memory, learning, consciousness, perception and all the thinking processes are located.

2. The *mid-brain* lies beneath the fore brain and links it with the medulla. This is where autonomic functions are localized: such basic sensations and reactions as hunger, thirst, anger, defensiveness, and flight, as well as vegetative control of the internal organs. Among its important constituents are the thalamus and hypothalamus.

3. The *hind brain* forms the link with the spinal cord and also includes the cerebellum. Here are located such vital functions as respiration, control of heart and blood circulation, as well as hiccuping, coughing, swallowing and vomiting. Nearly all the nerves from the brain connect with the medulla, which has a dense and complex network of synapses, the *formatio reticularis*, or reticular formation.

Figure 115 shows a few parts of the brain in a simple diagram.

The limbic system is an elongate structure, lying beneath the cerebral hemispheres (the two parts of the cerebral cortex) and forming a link with the deep parts of the brain (mid-brain). It consists in part of cortical tissue and in part of nervous structures belonging to the underlying nerve centres. The

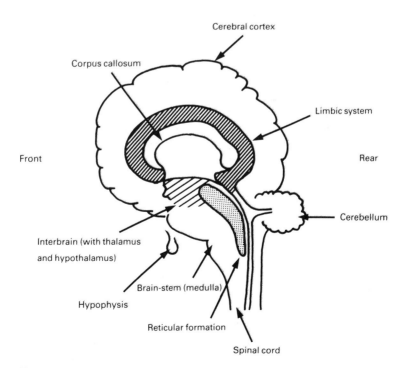

Figure 115. General plan of the layout of the brain.
A longitudinal section to one side of centre line; the hypophysis requires a median section.

limbic system is believed to play an important part in generating emotions and the physical excitement that goes with them. In addition, this system is involved in the direction of day-/night rhythms, appetite, sexual behaviour, motivation and excitements such as rage and fear.

If the lateral part of the limbic system (the hippocampus) is destroyed it results in serious deficiencies in short-term memory, although long-term memory remains intact, but we still do not know the exact location of short-term memory.

The formatio reticularis is an activation system essential to the processes of learning and memory, since it is only during waking hours that memories can be stored away. The more active the brain, the more learning and memorizing go on, and doubtless the higher the level of motivation.

Long-term memory

Long-term memory is very stable and resistant to both brain disturbance and electric shock. This fact has led to the conclusion that long-term memory must depend upon some form of intramolecular storage of stimuli, that is, to changes in the chemical substratum of the nerve cells. Many experiments suggest that this may be a durable engraving in the ribonucleic acid (RNA).

Fascinating experiments with planarians have shown that acquired behaviour is not lost when these flatworms are cut up. If a worm is trained to react in a particular way to a certain stimulus, and is then cut into two, both head and tail regenerate complete worms, and both of these retain the acquired behaviour! This led to the hypothesis that acquired behaviour is based upon changes in the RNA. It is well-known, of course, that the RNA provides the template for the synthesis of proteins during cell division; hence it is assumed that the cells of each cut half are replicated into the regenerated half complete with their 'marked RNA'. This hypothesis is confirmed by a further experiment: if either half is treated with ribonuclease, an enzyme that destroys RNA, then the acquired behaviour is lost, and the regenerated worm is no longer conditioned.

Many other experiments suggest that long-term memory is engraved upon the RNA, including several studies of learning in mammals, which were accompanied by an increase in the RNA in the brain. In one study, cited from Müller-Limmroth [192], rats were trained to use only the left forepaw to reach for food, and the corresponding centre in the brain showed an increase in RNA content. Information theory also gives interesting insights into memory processes. The number of nerve cells in the brain that are concerned with memory storage may be estimated at 10 billion (10^{10}) and assuming that all these cells are actively involved, *the storage capacity of the human memory may be as great as $10^8 - 10^{15}$ bits* [79]. This is an inconceivably large storage capacity.

A problem for human beings is to be able to 'recall' the stored information. Everyone knows the difficulties that older people have, and how sadly we say 'I can't call it to mind any more'. So far science can do nothing about this problem.

10.4. Sustained alertness (vigilance)

Jobs in industry and transport that call for sustained alertness are especially demanding mentally. Before we consider the actual problems of remaining alert, it is sensible first to discuss some of the activities that are often used as indicators of the level of mental efficiency.

Reaction time

Psychologists and ergonomists have particularly concerned themselves with *speed of reaction*: psychologists because a study of reaction times gives them an insight into mental problems, and ergonomists because reaction time can often be used as a way of assessing the ability to perform mental tasks. Reaction time means the interval between the receipt of a signal and the required response. According to Wargo [267] this time is made up of the following parts:

(*a*) conversion into a nerve impulse in the
sense organ: 1−38 ms
(*b*) transmission along a nerve to the cerebral
cortex: 2−100 ms
(*c*) central processing of signal: 70−300 ms
(*d*) transmission along a nerve to musculature: 10−20 ms
(*e*) latent time of response of muscle: 30−70 ms
Total time, say: 100−500 ms

It will be seen from this that a substantial part of the reaction time is taken up with mental processing of the signal in the brain. The considerable range in each of these times must be attributed to the wide variety of sense organs studied, varying lengths of sensory and motor pathways, and differences in the type of signal to be processed.

Simple reaction time

A simple reaction time is one that involves a simple signal, and one that is expected, which is answered by a simple motor reaction. The average time for such a reaction is 0·15−0·20 seconds. According to Swink [251] *simple reaction times* under various conditions of study, produced the following average figures:

(*a*) light signal: 0·24 s
(*b*) siren: 0·22 s
(*c*) electrical shock on skin: 0·21 s
(*d*) light signal + siren: 0·20 s
(*e*) all three at once: 0·18 s

Selective reaction time

If a variety of signals must each be answered with a different reaction, we speak of *the selective or choice reaction time*. For example, a signal may be green, red or yellow, and each must be answered by pressing a different key. These reaction times are substantially longer than simple reactions, because the decision requires the incoming signal to be processed by the brain, and the number of possible answers is an essential factor.

A summary by Damon *et al.* [51] suggests the following relationship between the number of possible answers and the consequent reaction time:

Number of answers		1	2	3	4	5	6	7	8	9	10
Approximate reaction time, in 1/100s of a second		20	35	40	45	50	55	60	60	65	65

Thus it can be assumed, to a first approximation, that selective reaction time increases linearly with the number of bits of information, up to a threshold of about 10 bits [116].

Anticipation

As we have said, both types of reaction time are shortened if the signals are anticipated. This is always the case in experimental work, whereas in everyday life most reactions are to unexpected stimuli.

In one investigation [268] short-hand typists were given a knob to be pressed whenever they heard a buzzing tone. This signal was sent out no more than once or twice a week, over a period of six months, in the middle of the typists' ordinary work. Reaction times under these conditions averaged 0·6 s, in place of the normal 0·2 s.

Movement times

So far no allowance has been made for the time taken up by the movement of the control, which can be especially important when controlling a vehicle, since it may be at least 0·3–0·5 s. The total elapsed time may then amount to 0·5–1 s. The importance of this factor to the driver of a vehicle reaches its extreme for the pilot of a supersonic aircraft. If the aircraft is flying at 1800 km/h it will travel a distance of 300 m during the total reaction time of 0·6 s.

Limits of mental load

It has long been known that thinking and other mental processes become less effective as time goes on. It is common experience that the longer one reads, the harder it becomes to take in the information, and the more often it is necessary to read a passage again before the words and sentences make sense. Who has not suffered the wandering of attention that occurs during a long and boring speech?

Bills' blocking theory

These simple, everyday observations have been amplified by the work of Bills [17], who demonstrated by psychological experiments that people cannot concentrate on a mental activity without respite. In practice there occurred at comparably frequent intervals interruptions in the processing of information, which Bills called 'blockings' or 'blocks'. The duration of a block extends to at least twice the length of the average processing time. Bills sees these blockings as *a kind of autonomic regulatory mechanism, which has the effect of allowing the level of mental effort to be as high as possible for as long as possible.*

Long-continued mental effort brings about more frequent and longer blocks, which can be seen as a fatigue symptom. Broadbent [29] studied fatigue symptoms of this kind in an experiment during which the test subjects had to remain continuously alert for the appearance of a weak optical signal of short duration, which was given only infrequently (15 times per hour). Broadbent found that various distractions (noise, heat, being deprived of sleep) made blocks appear more quickly, and more signals were missed. The author compared blocking with the closing reflex of the eyelids, which is an interruption of continuous perception of light, and which also becomes more frequent and lasts longer as fatigue increases.

Variability of heart rate

The variability of the heart rate has been used as an indicator of mental stress. In practice heart rate is not regular from one beat to the next, but constantly varies both up and down. The physiological term for this variation is *sinus arrhythmia*, which is mainly linked to the act of breathing. At each inspiration the heart rate rises, to slow down again during the following expiration. This arrhythmia seems to be directed, in the first instance, by the autonomic nervous system, with the vagus nerve playing an important role as 'pacemaker'. Several authors, among whom Kalsbeck [139] and O'Hanlon [206] may be mentioned, agree in finding that the variability of the heart rate is less during either physical or mental stress. O'Hanlon [206] found that when his research subjects relaxed their concentration, their heart rate became more variable, and he suggested that this rise in variability might be used as an indicator of the level of concentration. It may be said, in simple terms that:

(a) *A fall in variability of heart rate may be a sign of increasing concentration.*
(b) *A rise in variability may accompany any fall in concentration.*

Sustained concentration or vigilance

Sustained concentration is the ability to maintain a given level of alertness over a long period of time. Vigilance is another term for this ability, which was of topical importance during World War II; it was noticed that the frequency with which look-outs noticed U-boats on the radar screen diminished with the length of the period on watch. 50% of all sightings were reported during the first 30 minutes on watch; during subsequent 30-minute periods these fell to 23%, then to 16%, and finally to 10%. Obviously alertness became less, the longer the time on watch. This wartime experience led to many studies of alertness, which came to be called 'vigilance research'.

Missed signals   In an early investigation, which has become almost a classic, Mackworth [174] set up a number of research subjects in a quiet situation and made them watch an electric clock. Each rotation of the hand occupied one minute and was divided into 100 steps of 1/100 minute each. Occasionally the hand jumped two steps at once, and the observers were required to notice this and respond to it. The experiment lasted two hours, and during each 30-minute period there were 12 critical signals (i.e. double jumps of the hand) at irregular intervals. The results are shown in Figure 116.

These results confirm the experience of the radar observers detecting submarines referred to earlier: the number of missed signals increased as time went on. Following this pioneer work by Mackworth, there have been hundreds of other experiments to find out how the level of alertness is affected by different kinds of signals, critical and non-critical, by their number and arrangement, and by the duration of the experiment. Leplat [165] rightly criticized vigilance research, which dissipated itself in the study of too many variables, using

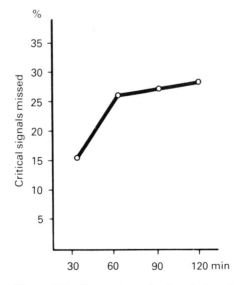

Figure 116. Frequency of missed signals during a test of vigilance lasting 2 hours by 25 sailors, as a percentage of that of the preceding 30-minute period.
Critical signal: a double jump of a pointer. After Mackworth [174].

questions that had too little relevance to practical conditions. In spite of these limitations it is possible, with hindsight, to pick out a number of results that have a practical value, and to formulate certain laws. Good reviews of the literature are given by Broadbent [29], Leplat [165], Schmidtke [232], Mackworth [173] and Davis and Tune [52].

Signal frequency and performance

Among all these results there is one that is significant for our later discussion of the problem of boredom. Several investigations have shown unmistakeably that the frequency with which signals are noticed rises when there are more such signals per unit time. Schmidtke [232] has shown that this rise continues up to an evidently optimal frequency of 100–300 critical signals per hour. If this limit is appreciably exceeded, then the observational performance falls off again. These results are set out in Figure 117.

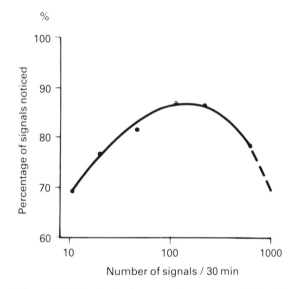

Figure 117. **Relation between frequency of signals requiring reaction and the observed performance.** *After Schmidtke* [232].

Schmidtke [232] concludes that the relation between the frequency per unit time of the critical signals presented and the observational performance has the shape of an inverted 'U'. Hence we may assume that too few signals leave the research subjects understressed, and conversely that excessively frequent signals make too heavy demands on them.

Jerison and Pickett [133] have observed in addition that the frequency of irrelevant signals also affects performance; the more there are, the poorer is alertness.

*Results from
vigilance
research*

The most important results to date from vigilance research may be summarized as follows:

1. Sustained alertness (measured by the number of signals noticed) decreases the longer the period of supervisory duty. The decline begins to be evident, as a rule, after the first 30 minutes.
2. Within certain limits the observational performance is *relatively improved*, if:
   the signals are more frequent;
   they are stronger;
   the subject is informed about his own performance;
   the signals are more distinct in shape or contrast.
3. Performance is *worse*, if:
   intervals between signals vary a great deal;
   the research subject has previously been under physical stress, or aroused from sleep;
   the research subject has performed under unfavourable conditions of noise, temperature, humidity, etc.
4. Many results of vigilance tests are strikingly similar to analogous research into reaction times. Both show a close relationship with the strength of stimulus, the interval between signals, and the current state of information available to the subject.

Theories

Psychologists, too, have not missed the opportunity to construct theories about vigilance! Thus we have between six and ten different theories to explain the varied observations made during concentration experiments. The three theories most worth considering may be mentioned briefly, as follows:

1. *Mackworth's internal inhibition theory* [173] invokes our knowledge of conditioned reflexes and explains the decline in performance as a consequence of inhibition because of the absence of 'rewards' and 'incentives'.
2. *The 'arousal' or activation theory* explains the fluctuations in concentration in terms of the level of activity of the cerebral cortex, as we know it from neurophysiology. According to this theory, a lack of external stimuli reduces the activity of the formatio reticularis, which in turn induces changes in the cerebral cortex, leading to increased drowsiness, and a lower level of concentration (see Chapter 11). Numerous experiments which have clarified different aspects of vigilance can be readily explained by the effect of the activation system on the functional state of the brain, and hence on observational efficiency.
3. *Broadbent's filter theory* [29] is based upon 'channel capacity', whereby a filtering mechanism allows only signals with certain characteristics to pass through in the stream of information. During the course of a task requiring alertness the filter becomes less and less discrimi-

nating, and allows irrelevant signals to pass. These overload the channel, exceeding its capacity and crowding out some of the relevant signals.

All these theories are still disputed, and none of the numerous experiments has been satisfactorily explained. Nowadays it seems doubtful whether sustained concentration can be considered as an isolated process, complete in itself. Perhaps that is why no one theory can fully explain it.

From a physiological point of view there is support for the belief that *sustained concentration depends on the functional state, i.e., the level of activity, of the cerebral cortex.* This is a dynamic equilibrium, controlled by a variety of influences, by stimuli which excite or damp down activity. No doubt the activation system of the formatio reticularis also plays a decisive role, but this is certainly not the only efficient factor. There are good reasons for assuming that other neurophysiological processes affect the level of activity and hence the degree of sustained concentration. Among these may be mentioned *habituation* (becoming accustomed to unimportant stimuli), *adaptation* (decline in intensity of stimuli transmitted by the sense organs), and finally *the part played by the limbic system of the brain in motivation and emotional reactions.* The processes of habituation and adaptation are discussed in more detail in Chapter 13.

The obvious conclusion is that the arousal (activation) theory best fits the observed facts of sustained concentration. If we consider all the many processes such as habituation, adaptation and the influence of the limbic system, there are few experimental results in vigilance research that cannot be accounted for by neurophysiological processes in the brain. Still, to say it is all a neurophysiological phenomenon is hardly to formulate a theory.

11. Fatigue

Terminology
'Fatigue' is a state that is familiar to all of us in everyday life. The term usually denotes a loss of efficiency and a disinclination for any kind of effort, but it is not a single, definite state. Nor does it become clearer if we define it more closely as physical fatigue, mental fatigue, and so on.

The term 'fatigue' has been used in so many different senses that its applications have become almost chaotic. A reasonable disinction is the common division into *muscular fatigue* and *general fatigue*.

The former is an acutely painful phenomenon, which arises in the overstressed muscles, and is localized there. General fatigue, in contrast, is a diffused sensation which is accompanied by feelings of indolence and disinclination for any kind of activity. These two forms of fatigue arise from completely different physiological processes, and must be discussed in separate sections.

11.1. Muscular fatigue

External symptoms
Figure 118 illustrates the external signs of muscular fatigue as they appear in an experiment with an isolated muscle from a frog. The muscle is stimulated electrically, causing it to contract and perform physical work by lifting a weight. After several seconds it is seen that:

(*a*) The height of lift decreases.
(*b*) Both contraction and relaxation become slower.
(*c*) The latency (interval between stimulation and beginning of the contraction) becomes longer.

Essentially the same result can be obtained using mammalian muscle. The performance of the muscle falls off with increasing strain until the stimulus no longer produces a response.

Human beings show this process, whether the nerve or muscle is stimulated electrically, or whether the research subject makes voluntary and rhythmical contractions of a muscle over a period.

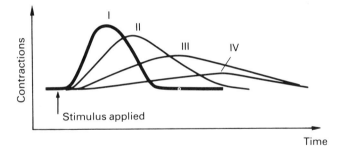

Figure 118. Physical manifestations of fatigue in an isolated muscle from a frog's leg.
I: Contraction and relaxation of a fresh muscle. II: The same, after moderate stress. III: The same, after heavy stress. IV: The same after most severe stress.

This phenomenon of reduced performance of a muscle after stress is called 'muscular fatigue' in physiology, and is characterized not only by reduced power, but also by slower movement. Herein lies the explanation of the impaired co-ordination, and increased liability to errors and accidents that accompany muscular fatigue.

Biochemical changes

We know that during muscular contraction chemical processes occur which, among other things, provide the energy necessary for mechanical effort. After contraction, while the muscle is relaxed and resting, the energy reserves are replenished. Thus both energy-releasing breakdown and energy-restoring synthesis are going on in a working muscle. If the demand for energy exceeds the powers of regeneration, the metabolic balance is upset, resulting in a loss of muscular performance.

After a muscle has been heavily stressed, its energy reserves (sugar and phosphorus compounds) are depleted, while waste products multiply, the most important of these being lactic acid and carbon dioxide. The muscular tissue becomes more acidic.

Electro-physiological phenomena

There are many references in the literature to the fact that even after a muscle has been exhausted by repeated voluntary contractions, it will still respond to an electrical stimulus applied to the skin, suggesting that this form of fatigue is a phenomenon of the central nervous system, i.e., of the brain, not of the muscle itself.

This interpretation cannot, however, be confirmed in every case. Many physiologists have made the same observation, that when the muscle itself is in an exhausted state it does not contract any more, even though further motor nerve impulses are visible on the electromyogram.

It seems that fatigue has now become a peripheral phenomenon, affecting the muscle fibres, as Scherrer [231] suggests. We must presume that one lot of people were looking for the first sign of fatigue, whereas the others were studying muscles that were already in a state of exhaustion.

Electromyo-
grams of fatigued
muscle

Comparing all the experimental results, we are led to the assumption that the central nervous system acts as a compensatory mechanism during the early stages of fatigue. Thus many studies with the electromyograph have shown that when a muscle is repeatedly stimulated its electrical activity increases, even though its contractions remain at the same level or decline. This must mean that more and more of its individual fibres are being stimulated into action (recruiting of motor units). The electromyogram in Figure 119 demonstrates this increasing electrical activity as fatigue increases.

These electromyographical phenomena have also been observed under practical conditions. For example, after 60−80 minute periods of perforating punched cards, increased electrical activity was recorded in the muscles of forearms and shoulders [283]. It is reasonable to conclude from this that the muscular fatigue that arises in industry is still at the stage where it can be compensated for by increased activity of the muscular control centres.

Another indicator of muscle fatigue seems to be the decrease in the frequency of the discharges of the muscular control centres. In fact, during fatiguing static contraction both an amplitude increase and a frequency decrease of the electrical activity is observed. Under rested conditions the mean frequency of the myoelectric signal may be twice as high as that of a fatigued muscle.

If we now turn to a state of exhaustion, which is certainly located in the muscles themselves, we find that this is

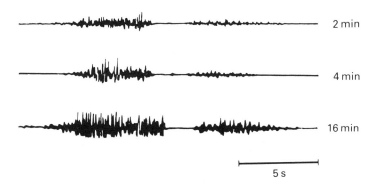

2 min

4 min

16 min

5 s

Figure 119. Three sections from the electromyogram of the extensor muscle of the upper arm, after it has been fatigued by a long series of contractions of equal strength.
From above downwards, the electromyogram after 2, 4 and 16 minutes' work. After Scherrer [231].

accompanied by a reduction in muscular strength. This can be partly compensated for initially by increasing discharges of the motor neurones.

11.2. General fatigue

A sensation of weariness

A major fatigue symptom is a general sensation of weariness. We feel ourselves inhibited, and our activities are impaired, if not actually crippled. We have no desire for either physical or mental effort; we feel heavy and drowsy.

A feeling of weariness is not unpleasant if we are able to rest, but it is distressing if we cannot allow ourselves to relax. It has long been realized, as a matter of simple observation, that weariness, like thirst, hunger and similar sensations, is one of nature's protective devices. Weariness discourages us from overstraining ourselves, and allows time for recuperative processes to take place.

Different kinds of fatigue

Leaving aside purely muscular fatigue, the following other kinds can be distinguished:

1. Visual fatigue, arising from overtiring the eyes.
2. General bodily fatigue: physical overloading of the entire organism.
3. Mental fatigue, induced by mental or intellectual work.
4. Nervous fatigue, caused by overstressing one part of the psychomotor system, as in skilled work.
5. Monotony of either occupation or surroundings.
6. Chronic fatigue, an accumulation of long-term effects.
7. Circadian or nyctemeral fatigue, part of the day—night rhythm, and initiating a period of sleep.

This classification of types of fatigue is based partly on the cause, and partly on the way in which the fatigue manifests itself, with the obvious corollary that the two should be linked as cause and effect. This should be particularly true for the different sensations of fatigue, which vary according to source. We believe, however, that there are certain regulatory processes in the brain that are common to fatigue of all kinds, and we shall now consider these in more detail.

Functional states

At any moment the human organism is in one particular functional state, somewhere between the extremes of sleep on the one hand, and a state of alarm on the other. Within this range there are a number of stages, as shown in the following summary:

Deep sleep	Light sleep, drowsy	Weary, hardly awake	Relaxed, resting	Fresh, alert	Very alert, stimulated	In state of alarm

Seen in this context, fatigue is a functional state which in one direction grades into sleep, and in the opposite direction into a relaxed, restful condition.

Before we go into the neurophysiological basis of this functional state it would be sensible to take a look at the most important method by which fatigue can be studied: the *electroencephalogram*, which records the electrical activity of the brain. When studying human subjects the electrodes are usually applied to the skin of the head, where they detect and register the waves of electrical potential in the cerebral cortex. The resulting encephalogram makes it possible to study the varying amplitudes and frequencies of these waves.

In greatly simplified form, the most important features recorded in an electroencephalogram are as follows:

1. The *alpha rhythms* comprise waves in the frequency band 8−12 Hz. Alpha waves are present during waking hours and are blocked by sensory impulses, so that a high alpha wave component indicates a relaxed condition and a reduced readiness to react to stimuli. A lower alpha component, coupled with a higher beta component indicates a more alert state.
2. The *theta rhythms* (4−7 Hz) are slow, long-period waves, which replace the alpha waves when we go to sleep. Sleep is characterized by a whole series of other phenomena, which cannot be discussed here.
3. The *delta rhythms* are also slow waves, of less than 4 Hz, which are present only during sleep.
4. *Desynchronization*, producing beta rhythms of 14−30 Hz, is an irregular electrical activity of very low amplitude. It occurs after the receipt of a sensory stimulus, and is the expression of an interruption of the synchronized activities of neurones which make up the alpha rhythm. (Synchronization is the simultaneous excitation of a large group of nerve cells.) *Desynchronization is the sign of a state of increased alertness, and it is also known as an arousal reaction.* Awaking and taking alarm are examples of this. It may be mentioned here that adrenalin (a stimulant derived from the adrenal medulla), which is released into the blood as a reaction to stress, acts by inducing desynchronization.
5. *Evoked potentials* are single fluctuations of potential which are induced by isolated sensory stimuli. They are located in that area of the cerebral cortex where the ascending nerve tract from that particular sense organ terminates (this has made it possible to plot a 'map' of the cortex, showing which area serves which sense organ). The first of 'primary' evoked potentials is followed by other 'diffuse secondary potentials', about which no more can be said here.

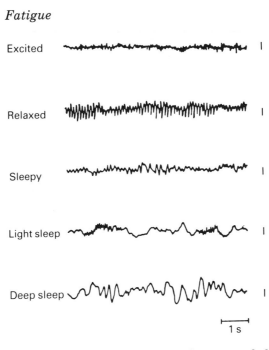

Excited

Relaxed

Sleepy

Light sleep

Deep sleep

1 s

Figure 120. Five sections from electroencephalograms, characteristic of various functional states.
The vertical lines indicate the scale for 1 µV. After Jasper [130].

Figure 120 shows five extracts from electroencephalograms, each characteristic of a different functional state of the body.

What has been said up to now gives the impression that for every functional state that we can describe by terms such as 'weary', 'lively', etc. there should be a distinctive pattern on the electroencephalogram. Unfortunately this is not the case. Those patterns shown in Figure 120 indicate no more than a very rough correspondence between the encephalogram trace and the underlying physiological state. Nevertheless it has been possible in the last 30 years to make some interesting discoveries about how various functional states are controlled, using electroencephalography in association with electrical stimuli and behaviour studies.

The reticular activating system

There is one very important nervous structure which controls to a large extent the functioning of the brain and, consequently, the whole organism: the so-called reticular formation of the medulla, which can increase or decrease the sensitivity of the cerebral cortex (see Figure 115). It is in the cerebral cortex that conscious awareness is localized, including the powers of perception, subjective feeling, reflection and will-power. Increasing the sensitivity of the cortex will therefore bring all the conscious functions to a higher level of alertness.

Thus it appears that the reticular formation controls the degree of alertness, including attention, and readiness for action. Its level of activity is very low during deep sleep, increases when sleep becomes shallow, and rises steeply on awakening. The higher the level of reticular activity, the higher the level of alertness, culminating in a state of alarm.

These reticular structures, which control the sensitivity of the conscious functions, are called *the ascending reticular activating system*, whereas those structures that increase the readiness of the skeletal musculature are the descending activating system. In what follows we shall be concerned solely with the arousal activating system.

The activating structures of the reticular formation do not, however, act on their own initiative. They themselves must be activated by afferent nerve stimuli, which come in the main from *two sources, the conscious sphere of the cerebral cortex and the sense organs*. What significance do these two have?

Feedback control   *Nervous tracts coming from the cerebral cortex* carry impulses from the conscious sphere into the reticular activating system. Such impulses arise for example, when an idea, or something noticed outside, seems ominous and calls for increased alertness. A closed circuit is then set up; the reticular activating system arouses the cerebral cortex and alerts conscious perception. If this results in any significant signals being received, then these stimuli send impulses back along the nerve tracts from the cortex to the reticular activating system. The result is a feedback system, analogous to that in many electronic devices.

Afferent sensory   *The other sources of stimuli to the reticular activating system*
system   *are the stream of afferent stimuli from the sense organs*; nerve fibres branch off from all the afferent nerve tracts and pass to the reticular activating system, and this sensory inflow has its effect upon the level of reticular activity. Every strong impulse that comes from the ear, the eye or from nerves conveying pain, can raise the level of reticular activity in a flash. The importance of this is obvious. Signals coming into the body from the outside world are passed over to the reticular activating system, and make this more active; this alerts the cerebral cortex and so ensures that the brain is ready to notice and act on what is happening outside the body. *This close linkage between the reticular activating system and the afferent sensory system is an essential prerequisite for conscious reaction to the external world.*

For example, suppose that a loud noise is heard. Any sudden noise stimulates the reticular activating system by way of the afferent sensory stream, and thereby increases the alertness of the cerebral cortex. There the noise is perceived and interpreted, perhaps resulting in precautionary action against possible danger. In that case the reticular activating system is operating as a distributor and amplifier of signals from the

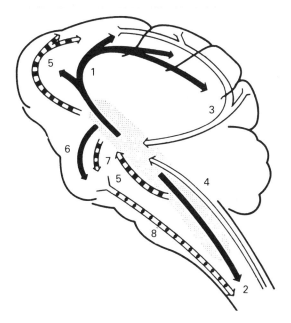

Figure 121. The activating and inhibitory systems in the brain.
Dotted area = reticular formation. 1 = ascending reticular activating system. 2 = descending activating system. 3 = pathways from the cerebral cortex. 4 = incoming sensory pathways. 5 = inhibitory (damping) system. 6 and 7 = links with vegetative (autonomic) centres. 8 = vegetative tracks leading to the autonomic nervous system of the internal organs.

outside world, and makes sure that the organism is alerted as much as is necessary for the preservation of its life.

Figure 121 shows diagrammatically the flow of stimuli from the cortex and the sense organs to the reticular activating system, as well as the presumed pathways in the cortex itself.

Limbic system and level of activity

The reticular activating system is not the only nervous organ that affects the state of readiness of the cerebral cortex, and through this, of the entire body. Yet another of the numerous functions of the limbic system must be mentioned: the part it plays in excitement, emotion and motivation. Figure 115 (Chapter 10) shows the anatomy of the limbic system and briefly enumerates its functions. The general level of alertness is greatly dependent upon the limbic centres for circadian rhythm (day−night periodicity), fear, rage and calmness, as well as for motivation.

Since the limbic system is involved in the generation of emotional states and the build-up of motivation, it seems obvious that *the activity level of the cerebral cortex, and through it of the entire organism, will be influenced by what goes on in the limbic system.* Links between the flow of afferent sensory impulses and the reticular activating system lead to the further assumption that the two systems have a reciprocal effect on each other.

Inhibiting system

These two activating systems are not alone, however, but there are other structures which clearly work in opposition to them. In fact many physiologists have conducted experiments showing the existence of nervous elements both in the inter-brain and in the medulla which have an *inhibitory* or *damping effect on the cerebral cortex*. These inhibiting pathways are shown in Figure 121. It is generally accepted that such inhibiting structures can finally induce sleep, but there is certainly no proof yet that they are related to states of fatigue.

Relations to vegetative functions

We know that there are close links between the autonomic (vegetative) nervous system, which controls the activities of the internal organs, and these activating and inhibitory systems. In fact any increase in stimulation of the reticular activating system is accompanied by a whole series of changes in the internal organs, of which we may mention the following:

Increase in heart rate.
Rise in blood pressure.
More sugar released by the liver.
Increased metabolism.

Ergotropic and...

This increased sensitivity spreads from the activating system to all parts of the body — brain, limbs and internal organs — until the entire organism is braced up for a period of high energy consumption, whether this is for work, for fighting, for flight, or whatever. Hess [111] coined the term '*ergotrope Einstellung*', or 'ergotropic tuning-up' for this process.

...trophotropic tuning-up

By an analogous process, increased activity of the inhibitory system lowers the heart rate and blood pressure, cuts back respiration and metabolism, and relaxes the muscles, while the digestive system works more vigorously to assimilate more energy. Hess' term for this is '*trophotrope Einstellung*' or 'trophotropic tuning-up', by which he means that this process accelerates the recuperative functions by assimilating more food and replacing lost energy.

Thus we see that the nervous structures in the inter-brain and brain stem (medulla) play a key role in the regulation and co-ordination of functional states of the body. *The brain as well as the internal organs are directed in a logical manner by these nerve centres*, as an example will show:

A fire alarm sounds in a factory. The acoustic signal raises the level of activity of the reticular formation in all the operatives within the danger zone. On the one hand this activates the cerebral cortex to a greater awareness of its surroundings, while alerting the internal organs ready for a greater consumption of energy and for physical performances.

Humoral

Humoral control means regulation by means of chemical substances which circulate in the body fluids. We have seen that the reticular activating system is not autonomous, but is dependent upon stimulation from the cerebral cortex as well as from afferent sensory signals. In addition to these purely nervous mechanisms there are also the humoral effects which have a part to play in tuning-up the activating system as well as regulating the sensitivity of the limbic system.

Adrenalin

We have already mentioned desynchronization, the effect produced by the performance hormones *adrenalin* and *noradrenalin*, and which must be interpreted as a raising of the degree of alertness and of the activation level of the body. Bonvallet *et al.* [21] deduced from what we now know that every time the activation level is raised by stimuli from outside the body there is a release of adrenalin, which in turn stimulates activity of the reticular formation. The increased activity from outside stimuli would soon fade away again if this hormone did not intervene to maintain it. In this theory, the reticular activating system is responsible for short-term reactions, while effects of longer duration depend upon the action of adrenalin.

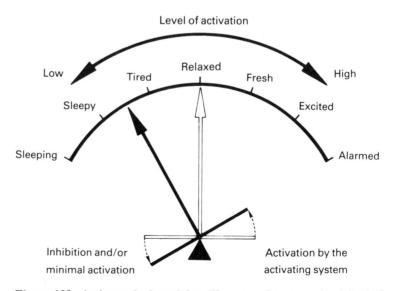

Figure 122. A theoretical model to illustrate the neurophysiological mechanism which regulates the functional state of the organism.
The level of activation of the cerebral cortex, the degree of readiness for action and the level of alertness all increase from left to right.

We can say, therefore, that *the level of activity of the reticular activating system depends upon:*

1. *The inflow of sensory stimuli.*
2. *Stimulation of the cerebral cortex.*
3. *Level of adrenalin.*

It must be pointed out that adrenalin and noradrenalin are also considered to be the stress hormones, since their secretion into the blood is greatly increased in stress situations. This will be discussed in Chapter 12.

Our concept can be summarized in different terms, as follows: the level of readiness to act lies somewhere between the two extremes of sleep and the highest state of alarm. Control mechanisms in the medulla and the inter-brain regulate this and adjust it to meet the momentary demands of the organism. If external influences are dominant, then the activating systems prevail; the person feels keyed up, even in a state of alarm, and is ready for action both physically and mentally. If, however, inhibiting influences from inside the body predominate, then the damping system prevails; the person feels sluggish, drowsy and lethargic.

Figure 122 compares the neurophysiological model with a balance, in which the activating system is regulating the general functional state of the central nervous system.

11.3. Fatigue in industrial practice

Causes of general fatigue

We know from everyday experience that fatigue has many different causes, the most important of which can be seen in Figure 123. This illustrates that the degree of fatigue is an aggregate of all the different stresses of the day. Visualize this diagram as a barrel partly filled with water. The recuperative rest periods would be the outflow from the barrel. To make sure that the barrel does not overflow we must ensure that inflow and outflow are of the same order of magnitude. In other words, to maintain health and efficiency the recuperative processes must cancel out the stresses. Recuperation takes place mainly during night-time sleep, but free periods during the day and all kinds of pauses during the work also make their contribution.

It must be emphasized that stress and recuperation must balance out over the 24-hour cycle, and that neither of these should be carried over to the next day. If rest is unavoidably postponed until the following evening, this can be done only at the expense of well-being and efficiency.

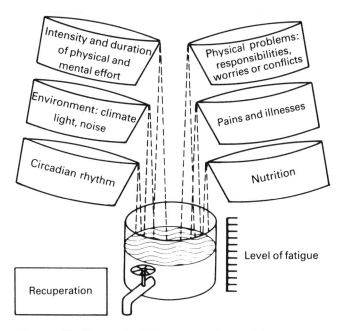

Figure 123. Theoretical diagram of the combined effect of everyday causes of fatigue, and the recuperation necessary to offset them.
The total stresses must be balanced by the total recuperation within the 24-hour cycle.

Symptoms of fatigue

Fatigue symptoms are both subjective and objective, the most important being:

1. Subjective feelings of weariness, somnolence, faintness and distaste for work.
2. Sluggish thinking.
3. Reduced alertness.
4. Poor and slow perception.
5. Unwillingness to work.
6. Decline in both bodily and mental performance.

Some of these symptoms result in a measurable drop in bodily and mental efficiency.

Clinical or chronic fatigue

Some of the fatigue states that arise in industrial practice are of a chronic nature. These are conditions that are brought about not by a *single* instance of overstrain, but by stresses which recur every day over long periods. Since conditions such as these are usually also accompanied by signs of ill health, this may correctly be called *clinical or chronic fatigue*.

Under these conditions the symptoms not only occur during the period of stress or immediately afterwards, but are more or less latent all the time. Tired feelings are often present on waking up in the morning, before work has even begun. This form of fatigue is often accompanied by feelings of distaste, which have an emotional origin. Such people often show the following symptoms:

1. Increased psychic instability (quarrelsomeness and associated behaviour).
2. Fits of depression (baseless worries).
3. General weakening of drive and unwillingness to work.
4. Increased liability to illness.

These ailments are mostly vague and come under the heading of psychosomatic complaints. This term is applied to functional disturbances of the internal organs or the circulation which are judged to be external manifestations of psychological conflicts and difficulties. Some of the commoner of these symptoms are:

Headaches.
Giddiness.
Loss of sleep.
Irregular heart-beat.
Sudden sweating fits.
Loss of appetite.
Digestive troubles (stomach pains, diarrhoea, constipation).

More ailments mean more absences from work, especially short absences, indicating that the cause of the absenteeism is the need for more rest.

People who have psychological problems and difficulties easily fall into a state of chronic fatigue, and it is often difficult to disentangle their mental from their physical problems. In practice, cause and effect are hard to distinguish in cases of clinical fatigue. This may be caused by dislike of the occupation, the immediate task, or the workplace, or conversely may itself be the cause of maladjustment to work or surroundings.

11.4. Measuring fatigue

Why measure fatigue?

The science of ergonomics is just as interested in the quantitative measurement of fatigue as is industry itself. Science studies the relationships between the fatigue and the level of stress, or more accurately fatigue and the output of work; otherwise, the reaction of the human body to different stresses can be measured in order to develop ways of improving work and making it less laborious. The question finally asked by industry is often simply whether the working conditions make excessive demands on the operatives, or whether the stresses involved are physiologically acceptable.

We are measuring only 'indicators' of fatigue	Discussion of measuring methods is subject to one serious limitation: *to date there is no way of directly measuring the extent of the fatigue itself*. There is no absolute measure of fatigue, comparable to that of energy consumption, expressed in kilojoules. All the experimental work carried out so far *has merely measured certain manifestations or 'indicators' of fatigue*.

Methods of measurement

Currently used methods fall into six groups:

1. *Quality and quantity of work performed.*
2. *Recording of subjective impressions of fatigue.*
3. *Electroencephalography* (EEG).
4. *Measuring subjective frequency of flicker-fusion of eyes.*
5. *Psychomotor tests.*
6. *Mental tests.*

Measurements such as these are frequently taken *before*, *during* and *after* the task is performed, and the extent of fatigue is deduced from these. As a rule the result has only relative significance, since it gives a value to be compared with that of a fresh subject, or at least with that of a 'control' who is not under stress. Even today we have no way of measuring fatigue in absolute terms.

Correlation with subjective feelings

More recently it has been the practice to study a combination of several indicators so as to make interpretation of the results more reliable. It is particularly important that subjective feelings of fatigue should also be taken into account. *A measurement of physical factors needs to be backed up by subjective feelings before it can be correctly assessed as indicating a state of fatigue*. The five indicators listed above will now be briefly discussed.

Quality and quantity of output

The quality and quantity of output is sometimes used as an indirect way of measuring industrial fatigue. Quantity of output can be expressed as number of items processed, as time taken per item, or conversely as the number of operations performed per unit time. Fatigue and rate of production are certainly interrelated to some extent, but the latter cannot be used as a direct measure of the former because there are many other factors to be taken into account: production targets, social factors and psychological attitudes to the work.

Sometimes fatigue needs to be considered in relation to the *quality* of the output (bad workmanship, faulty products, outright rejects), or to the frequency of accidents, once again with the reservation that fatigue is not the only causal factor.

Subjective feelings

Special questionnaires have been used to *assess subjective feelings*. One of these is the so-called bipolar questionnaire, where the subject is required to place a mark between the opposed items according to his feelings. An example of such a bipolar questionnaire shows the following opposing items:

Fresh   ———————— Weary
Sleepy   ———————— Wide awake
Vigorous   ———————— Exhausted
Weak   ———————— Strong
Energetic   ———————— Apathetic
Dull, indifferent   ———————— Ready for action
Interested   ———————— Bored
Attentive   ———————— Absent-minded

Other authors have suggested more complicated questionnaires, among which may be mentioned the comparison process of Pearson [208] and the very comprehensive questionnaire used by Nitsch [204].

The electro-encephalograph

The electroencephalograph is particularly suitable for standardized research in the laboratory, where variations in the trace in the sense of increasing synchronization (increase of alpha and theta rhythms, reduction of beta waves) are interpreted as indicating states of weariness and sleepiness (see Figure 120).

The techniques of detecting and recording have been improved recently, so that the electroencephalograph can now be used successfully to monitor sedentary activities, such as driving a vehicle [207, 284].

Flicker-fusion frequency of the eye

During the last 25 years the *flicker-fusion frequency of the eye* has been increasingly used as an indicator of the degree of fatigue. This is measured as follows. The research subject is exposed to a flickering lamp, and its frequency is increased until the flickers appear to fuse into a continuous light. The frequency at which this occurs is called the subjective flicker-fusion frequency. The source of light should have an area that subtends an angle of $1-2°$ at the eye, and should be so placed that it does not call for any optical accommodation. Several improvements of the measuring equipment and of the procedure have come into use [213, 80].

Lowering of the flicker-fusion frequency

It has been observed that reductions in the flicker-fusion frequency of $0·5-6$ Hz take place after mental stress, as well as under various industrial stresses. Yet a survey of the literature shows that not every kind of stress brings about such a reduction. Experience to date can be summarized thus, to a first approximation:

1. *A distinct lowering* of the flicker-fusion frequency can be expected during unbroken high level mental stress. Examples are calculating in one's head; working as a

telephonist; piloting an aircraft; carrying out demanding visual work.

We shall see later (in Chapter 13) that dull, repetitive and monotonous situations produce a distinct lowering of the flicker-fusion frequency.

2. *Either little lowering* or *none at all* result from work that requires only moderate mental effort, and which allows the operative comparative freedom of action, or which involves physical effort. Examples are office work, sorting jobs, and repetitive work at a moderate level.

During several studies the reduction in flicker-fusion frequency was accompanied by parallel changes in other signs of fatigue, notably by increased feelings of weariness and sleepiness. An example, is an experiment by Annette Weber and others [270] in which eight subjects in each of three tests were given a dose of 5 mg of diazepam (Valium) to produce 'pharmacological fatigue'. In a control experiment the same persons were given a placebo (a similar tablet without the drug). Measurements were taken of both the flicker-fusion frequency and the subjective feelings of fatigue, the latter by using a bipolar questionnaire.

The results, set out in Figure 124 show that after the administration of Valium there was a distinct lowering of the flicker-fusion frequency, amounting to about 2 Hz on the average. Simultaneous records of subjective feelings by the bipolar

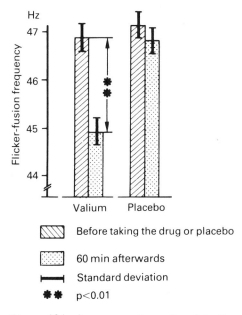

Figure 124. Average value of subjective flicker-fusion frequency before and after taking 5 mg diazepam (Valium), or a placebo. *Twenty-four experiments, each with eight test subjects. After Weber* et al. [270].

questionnaires showed that in 10 out of 15 of the opposing items of feelings there was a significant shift in the direction indicating increased fatigue.

This result shows a good correlation between flicker-fusion frequency and subjective feelings of fatigue. The statistics show that the same persons who showed a marked lowering of flicker-fusion frequency also exhibited considerable subjective fatigue.

In spite of these and other similar observations we can still say nothing of general application about the relationship between flicker-fusion frequency and subjective feelings of fatigue, since the individual correlations recorded here have not been confirmed by other experiments.

Nevertheless, these experiments collectively have encouraged most authors to interpret a *lowering of the flicker-fusion frequency as a sign of fatigue*.

In recent years the subjective flicker-fusion frequency has been less used in fatigue studies. The main reason for this might be some controversial results, the impossibility of obtaining a quantitative measure of fatigue and the rare correlations with other symptoms of fatigue.

Psychomotor tests

Psychomotor tests measure functions that involve perception, interpretation and motor reactions. Tests very often used are the following:

Simple and selective reaction times.
Tests involving touching or pricking squares in a grid.
Tests of skill.
Driving tests under simulated conditions.
Typing
Tachistoscopic tests (see Chapter 9) to measure performance involving perception.

Reservations

In tests like these it is also assumed that a decrease in performance can be taken as a sign of a state of fatigue. Since, however, ability to perform a psychomotor test is dependent on other factors such as, for example, motivation, it is sometimes doubtful whether a general state of fatigue is really the main cause.

A further disadvantage of psychomotor tests arises from the fact that often the test itself makes heavy demands on the subject, thereby raising the level of excitability. In view of what we have said previously, it is very likely that such tests will cause some kind of cerebral activity, which may at least temporarily mask any possible signs of fatigue.

Performance of mental tests

Performance of mental tests often involves:

Arithmetic problems.
Tests of concentration (e.g., crossing-out tests).

Estimation tests (e.g., estimation of time intervals).
Memory tests.

The same reservation must be made as for psychomotor tests: the test itself may excite the interest of the person being examined and so cancel out any signs of fatigue. Other disturbing factors are the effects of training and experience, and, if the test is protracted, fatigue brought on by the test itself.

Fatigue has been investigated in many field studies, carried out under industrial conditions, in traffic and in schools. As a rule, their significance is limited to a particular problem in a particular setting, and almost nothing can be deduced from them that is of wider application or would lead to generalization about the relationship between stress and fatigue.

Nevertheless, a few indications about some of the field studies shall be given here; for more details the reader is referred to the literature.

Traffic fatigue   A particular importance attaches to studies of fatigue in traffic, because it is reasonable to suppose that fatigue is an important contributory factor in mistakes and accidents. As early as 1936 Ryan and Warner [226] concluded from an extensive study on truck drivers that long periods of driving led to a reduced ability to discriminate between certain sensory impressions and to a loss of efficiency in some motor functions. Several authors have shown unmistakeably that about four hours of continuous driving is enough to bring on a distinct lowering of the level of alertness and thereby increase the risk of accidents. More recently electroencephalography as well as variability of heart rate have been used to measure fatigue in road traffic [206]. Most studies in this field have been described in detail by Lecret [161].

Fatigue, for a long time a problem among telephone operators, has been the subject of a number of studies [82]. Other investigations have been performed on employees of postal and railway services [89] and air traffic controllers [83, 91].

The investigations among bus drivers [206] and air traffic controllers showed interesting parallels. Both occupations call for sustained vigilance; in both cases the first signs of reduced efficiency appear after about four hours, and this becomes very marked after seven or eight hours. *This decline is a symptom of a fatigue state, which shows itself in both groups as:*

Subjective fatigue.
Fall in flicker-fusion frequency.
Decrease in psychomotor efficiency.
Decrease in driving precision.
More irregular heart beat.

Fall in heart rate.
Rise in alpha waves in the electroencephalogram.

It is difficult to deny the obvious assumption that all these symptoms are expressions of a decline in the level of activation (arousal) of the central nervous system.

One final conclusion is inevitable. Occupations that demand sustained vigilance must be so planned, with working periods and rest periods, that the risk of accidents is not increased through fatigue of the operators. The research work detailed above shows that these conditions are not being fulfilled at the present time.

12. Occupational stress

12.1. What is stress?

The original definition of stress by Selye

The term stress was introduced by the Canadian Selye [249] after World War II in the field of medicine. *He defined stress as the reaction of the organism to a threatening situation*, and distinguished between the stressor as the external cause and stress as the reaction of the human body. Selye had discovered that stress was essentially a chain of neuroendocrine mechanisms, beginning with an excitation in the brain stem, followed by an increased secretion of some hormones from the adrenal gland, especially of *adrenalin* and *noradrenalin*, known as 'performance hormones', since they keep the whole organism in a state of heightened alertness. These performance hormones, also called *catecholamines*, can be determined in the urine, and this is still a possible way of determining stress.

Physiological reactions

These hormones were already mentioned in Chapter 11 in connection with the role of the activating system, mainly located in the reticular formation. It was said that an increase in stimulation of the reticular formation is accompanied by an increase in heart rate and blood pressure as well as by an increased sugar level and metabolism. This reaction is called the 'ergotropic tuning-up' and is essentially identical with the basic mechanisms of the stress reaction. They reflect an intensified readiness to defend life, including fighting, fleeing or other physical achievements. But Selye also observed that

Health troubles

this emotional state, resulting from the feeling of being threatened, was responsible for the adverse effects of stress. In fact, long-lasting or recurrent stress situations can be detrimental to health by inducing functional troubles, particularly in the gastro-intestinal or in the cardiovascular systems. These effects are psychosomatic disturbances which, in the long run, can turn into organic illnesses. The most common forms of stress diseases are probably gastro-intestinal disorders, which can lead to gastric or duodenal ulcers. Selye explained the adverse stress effects on health as a maladaptation of the organism to stress.

Is stress always harmful?

Thus it is obvious that stress is part and parcel of our life; it is a necessary condition for all living creatures to react to threatening situations in an appropriate way. A life without stressors and stress would not only be unnatural, but also boring. Stress cannot be divorced from life, just as birth, death, food and love are inseparable.

Paracelsus, a medical doctor of the early 16th century, said that *the dose determines whether a compound is toxic or not (Dosis sola facit venenum)*. The same holds true for stress: the amount determines whether stress will have adverse effects on health or whether it will increase human ability to cope with life. Where the borders between normal physiological and pathological stress are to be drawn is still an open question. Only one thing is certain: this border varies from one individual to another. One person can bear a great amount of stress all his life; another suffers immensely and will sooner or later be overwhelmed by it.

The more the term 'stress' was used, the more it became a myth. Eventually the word was used for nearly every kind of pressure on people. In the last two decades, however, psychologists and social scientists have done detailed research on the phenomenon of stress and formed a much clearer conception of it, particularly with respect to occupational stress. Important contributions to the present concept of stress were developed by Lazarus [160], McGrath [184], Harrison [103], Cox [50] and Caplan [39].

Occupational stress

The emotional state (or mood) which results from a discrepancy between the level of demand and the person's ability to cope defines occupational stress; it is thus a subjective phenomenon and exists in people's recognition of their inability to cope with the demands of the work situation.

A stressful situation is a negative emotional experience which can be associated with unpleasant feelings of anxiety, tension, depression, anger, fatigue, lack of vigour and confusion. These feelings characterize the state of mood, often studied by specially designed questionnaires (Profile of Mood States: POMS).

Person−environment fit

Research on occupational stressors has come up with the concept of the *person−environment fit*. *The basic assumption is that the degree of fit between the characteristics of a person and the environment can determine the well-being and performance of workers.* Environment is used here in its largest sense and includes the social as well as the physical environment. Some authors distinguish the fit between the person's needs and their satisfaction through the job environment, others refer to the fit between the demands of the job environment and the relevant worker's ability to meet those demands.

Stressors in the work environment

Surveys as well as theoretical considerations suggest that the following conditions may become stressors in work environments:

1. *Job control* is the worker's participation in determining the work routine, including control over temporal aspects and supervising work processes. Several studies suggest that lack of control may produce emotional and physiological strain.
2. *Social support* means assistance through supervisors and peers. Social support seems to reduce adverse effects of stress. On the other hand, a lack of social support increases the load of stressors.
3. *Job distress or dissatisfaction* is mainly related to job content and work load. It is the perceived stress in job and career.
4. *Task and performance demands* are characterized by the workload, including demands upon attention. Deadlines may be a major stressor, too.
5. *Job security* today refers mainly to the threat of unemployment. Nowadays many office workers worry about being made redundant. Important is the recognition of the availability of similar or alternative employment and of future needs for their professional skills.
6. *Responsibility* for the lives and the well-being of other people may be a heavy mental burden. It seems that jobs with great responsibility are associated with an increased proneness to peptic ulcers and high blood pressure. Responsibility in itself is perhaps not the key stressor. The crucial question is rather whether the amount of responsibility exceeds one's resources.
7. *Physical environmental problems* include noise, poor lighting, indoor climate or small, enclosed offices.
8. *Complexity* is defined as the number of different demands involved in a job. Repetitive and monotonous work is often characterized by a lack of complexity, which seems to be an important predictor of job dissatisfaction. On the other hand, too high complexity can arouse feelings of incompetence and lead to emotional strain.

Any individual may experience a number of other stressors, and this list could be easily extended. But the 8 stressors mentioned above are certainly those which are often taken into account by social scientists when preparing their questionnaires to evaluate people's experience with occupational stress.

12.2. The measurement of stress

Cox [50] writes: 'Stress, as an individual psychological state, is to do with the way the person sees and then experiences the (work) environment. Because of the nature of the beast, there can be no direct physiological measures of stress. The measurement of stress at work must focus on the individual's psychological state. A first step is thus to ask the person about their emotional experiences or mood in relation to the situation at work. This means using state-dependent subjective data.'

Questionnaire surveys

Nowadays all field studies on occupational stress are based on extensive questionnaire surveys on working conditions, workers' health and well-being, potential stressors, job satisfaction and states of mood. Many authors use scales that have become standardized and widely used instruments for which normative data are available.

A popular method of evaluating psychological response criteria today is the use of mood checklists. These procedures serve to gauge the worker's feelings. The checklist of Mackay *et al.* [172] for instance, distinguishes between stress and arousal. To give an example, after a prolonged and monotonous repetitive task significant increases in self-reported stress were found together with significant decreases in self-reported arousal.

Another approach is that of psychosocial questionnaires which evaluate perceptions and feelings about the job situation, including job satisfaction, perception of work load, work pace, career opportunities, supervisory style and organizational environment. One of the most often cited and widely used questionnaires was developed by Caplan *et al.* [39]; it was used to measure various psychosocial aspects of 23 different occupations and is even able to make a distinction between the jobs and these psychosocial factors.

Finally, it must be pointed out that several authors, mainly Swedish scientists, also combined physiological parameters of stress with questionnaires, such as the excretion of catecholamines (adrenalin and noradrenalin) in the urine, heart rate and blood pressure. These measurements are certainly interesting correlates to the questionnaire survey procedures; they will be discussed in Chapter 13.

12.3 Alleged stress among VDT operators

General experience

Anecdotal reports as well as general experience indicate that the introduction of VDTs into offices also creates some psychological problems. In some cases the new technology imposes a performance increase and therefore a greater workload. To illustrate this, an example of payment transfers in a bank: without a computer, about 30 payment transfers were settled per employee per working day, but with the aid of a VDT the same employee handled about 300 transfers per day. On the other hand, some new VDT jobs became more and more repetitive and monotonous, especially data-entry jobs.

Some worried

Some employees worried; they were afraid of new technologies, automation and unemployment. This rather complex and only vaguely recognizable situation sometimes gave rise to a general negative attitude towards the new VDT job.

Some had fun

However, a contrary reaction was also often observed. In fact, some employees were proud of being included in the new information technology and looked forward to interacting with a computer. Those jobs that required creative participation from the operators were rated interesting. Many managers observed that clerical employees showed some resistance to word processing procedures in the beginning, but after a few weeks clearly preferred the new secretarial work with VDTs to their former job conditions.

In general, it seems that many psychological problems were more acute when VDTs were first introduced and that they are becoming less urgent as time goes by.

The Swedish study on stress at VDTs

One of the first surveys focusing on stress and job satisfaction at VDT workplaces was conducted in 1977 by Gunn Johansson and Gunnar Aronsson [135] on 95 employees in a large insurance company.

The questionnaire survey revealed rather positive attitudes to the job. But a certain anxiety characterized their opinion about the computerization: the 'data-entry' group had slightly higher catecholamine levels than the control group. The most striking result was observed during a temporary breakdown of the computer system: adrenalin, blood pressure and heart rate were raised, while at the same time the subjects felt more irritated, tired, rushed and bored. The authors concluded that *computer breakdowns are an important cause of mental strain for persons with extensive VDT work*. In fact, an interruption meant that the VDT operators were condemned to idleness while their own work piled up, which presumably increased the next day's work load. The authors believe that stress and strain at VDTs may be partly counteracted at the technological and the organizational level: by reducing the duration and frequency of breakdowns, by shortening the response times in the system and by redistributing unavoidable but monotonous data-entry work.

Many other studies have been carried out on alleged stress among VDT operators; they are discussed in detail in the book *Ergonomics of Computerized Offices* [85]. Here only a summary of 3 selected surveys shall be given.

The French survey

The French ergonomists Elias and Cail [66] observed an increased incidence of gastro-intestinal symptoms, anxiety, irritation and sleep disturbances among a 'data-acquisition' group.

*The NIOSH
study by Smith*

Two field studies were carried out in the USA by Smith *et al.*
[245, 246]. It was found that the VDT operators as well as the
control subjects were subjected to a large number of psychosocial stressors. In general, the VDT operators reported more
psychosocial stress than the control groups. The authors
concluded that the job content may be an important factor for
increased occupational stress and health complaints. The
results suggested that the use of VDTs is not the only factor
contributing to operator stress, but that job content, too,
plays an important role in this matter.

*The NIOSH
study by Sauter*

Sauter *et al.* [229, 230] also conducted a large survey, focusing
on job-attitudinal, affective and somatic manifestations of
stress among VDT office workers. None of the well-being
indices, related to job stressors and moods, disclosed a strong
indication of increased strain for the VDT group. The VDT
users had a higher incidence of unfavourable working conditions and rated their working place environment less pleasant
and their chairs less comfortable than the control subjects.
The authors summarize their findings as follows: "Other than
a tenuous indication of increased eye strain and reduced
psychological disturbances among VDT users, the two groups
were largely undifferentiated on job-attitudinal, affective, and
somatic manifestations of stress".

Most of the psychosocial surveys on stress among VDT
operators have been criticized, mainly because the designs of
the studies have not allowed conclusiveness. In spite of some
methodological deficiencies it is possible to draw the following
tentative conclusions:

1. *Generally speaking, clerical VDT operators do not show
   symptoms of excessive stress.*
2. *There is one important exception, though: some VDT
   operators, engaged in very fragmented, repetitive and
   monotonous jobs, such as data-entry or data acquisition,
   experience on average stronger psychosocial stressors,
   report low job satisfaction and indicate a higher frequency
   of mood changes for the worse, as well as gastro-intestinal
   or other psychosomatic troubles.*
3. *The fact that other studies, including repetitive VDT jobs,
   did not reveal more psychosocial stressors or symptoms of
   stress than control groups, leads to the conclusion that it is
   not the work with the VDT as such but the poor work
   structure of some specific repetitive jobs which is responsible for the observed adverse effects.*
4. *Apart from repetitive and monotonous jobs, VDT
   operators are, on the whole, satisfied with their work and
   consider a VDT an efficient tool.*

13. Boredom

13.1. Causes

*A monotonous environment is one that is lacking in stimuli,
and, at least in Western countries, the reaction of an individual to monotony is called boredom. Boredom as conceived by
Western scientists is a complex mental state characterized by
symptoms of decreased activation of higher nervous centres,
with concomitant feelings of weariness, lethargy and diminished alertness.*

Boring situations are common in industry, travel and com-
merce. They can be found, for example, at a control desk if
there are too few signals to which the operator needs to
respond. An engine driver can be in a similar situation if
signals are too far apart. An example of a monotonous job is
being in charge of a stamping press and having to carry out
exactly the same operation 10–30 times per minute, for hours,
days and years on end. Occupations such as this are *repetitive*
as well as monotonous and boring.

During the last 20 years many psychologists, as well as a
few physiologists, have concerned themselves with the problem of boredom. The psychologists have mainly described the
external causes of boredom and the behaviour of persons
suffering from it. The physiologists have concerned themselves more with the nervous mechanisms of boredom and
related these to the measurable indicators of this condition.
Although boredom is a *single* condition, it will be considered
from these two opposite points of view.

External causes
Experience shows that the following circumstances give rise
to feelings of boredom:

1. *Prolonged repetitive work that is not very difficult, yet
   which does not allow the operator to think about other
   things entirely.*
2. *Prolonged, monotonous supervisory work, which calls for
   continuous vigilance.*

The decisive factor in these situations is obviously that
there are not enough matters that call for action.

Observations in industry have shown that certain conditions make boredom more likely. Examples are a very brief cycle of operations and few opportunities for bodily movements. Others are dimly lit or warm workrooms, and solitary working without contacts with fellow workers.

Personal factors enhancing boredom

Personal factors have a considerable impact on the incidence of boredom, or, put another way, *on the ability to withstand boredom*. Proneness to boredom is higher for

1. People in a state of fatigue.
2. Not-adapted night workers.
3. People with low motivation and little interest.
4. People with a high level of education, knowledge and ability.
5. Keen people, who are eager for a demanding job.

Conversely, the following are very resistant to boredom:

1. People who are fresh and alert.
2. People who are still learning (i.e., a learning driver has no time to be bored).
3. People who are content with the job because it suits their abilities.

Extroversion

Several pieces of research concur in suggesting that extrovert people are very susceptible to boredom. On the other hand, the opinion often expressed that women are more resistant to boredom than men is scientifically questionable; equally, the supposed relationship between intelligence and susceptibility to boredom is still disputed.

Satiation

Many authors draw a distinction between boredom itself and its emotional manifestations, which they call 'satiation'. This means a state of irritation and aversion to activity which is provoking boredom. The person feels that he or she has 'had enough'. This is a state of actual conflict between a feeling of duty to work and the desire to have done with it, which puts the person involved under increasing internal tension.

Job satisfaction

A decline in work satisfaction can be looked upon as a precursor of mental satiation. Several studies have shown that in practice work satisfaction is lower where monotonous, repetitive work is concerned, than with jobs that allow a greater freedom of action. As an example of such a survey the results of Wyatt and Marriott [278] may be recalled. These authors questioned 340 workers in one car factory (Factory A) and 217 in another (Factory B) about their attitudes to their work. The authors compiled an 'Index of Satisfaction' from certain replies, and from self-assessments carried out by the workers themselves. Table 22 summarizes the most important of these:

Table 22. Assessment of work and an index of satisfaction by 557 workers in two car factories which had the work organized differently.
After Wyatt and Marriott [278].

Work	Factory	Production line (motorized)	Assembly line (non-motorized)	Free assembly
"Interesting"	A	35%	56%	67%
	B	34%	57%	94%
"Boring"	A	54%	42%	39%
	B	55%	37%	28%
Index of job satisfaction	A	0·53	0·92	0·96
	B	0·57	1·00	1·17

The workers at the two factories agreed in rating working on the motorized production line as more rarely 'interesting' and more often 'boring' than work at either the non-motorized assembly line or the free assembly. The index of job satisfaction was comparable for both factories. The same authors also compared this index with the length of shift of a selection of workers, and found that the shorter the shift, the greater the job satisfaction.

13.2. The physiology of boredom

We have seen already that situations in which stimuli are few, or lacking in variety, induce a state of boredom, recognizable by weariness and somnolence, as well as by a decline in alertness.

Neuro-physiological basis

This state of affairs is not difficult to explain in neurophysiological terms: *When stimuli are few the stream of sensory impulses dries up, bringing about a reduction in the level of activation of the cerebrum, and thereby of the functional state of the body as a whole.*

Besides the reduced sensory inflow during quiet conditions there are two other physiological processes that should be noted, because they, too, are responsible for the decline in the level of stimulation, particularly in situations where the existing stimuli vary very little. These are *adaptation* and *habituation.*

Adaptation

Most sense organs have the peculiarity that *under a prolonged, steady stimulus the discharge from the receptor organ declines.* Obviously one function of this is to protect the CNS against prolonged overloading with impulses from the peripheral sense organs. Hence the term *adaptation*: the stream of sensory impulses is adapted to the needs of the organism.

In principle, all sense organs have this power of adaptation, even though they differ in the extent and speed of it.

Adaptation is particularly well-developed in the sensitivity of the skin to pressure (we soon get used to wearing a wrist watch), the stretch receptors of the muscles, and the photo-receptors of the eyes.

Adaptation is not confined to the receptors of the peripheral sense organs, but also occurs in the synapses joining one nerve fibre to another.

What significance has adaptation in the problem of boredom? As we have said, the sense organs adapt themselves to external circumstances in such a way that they respond mainly to changes in stimuli and are relatively insensitive to a sustained level. Under uniform stimulation, therefore, the activating structures in the brain do not pass on any stimulus to those organs (reticular and limbic activating systems) responsible for the general level of activation of the body.

Habituation

Habituation can be regarded as adaptation on a higher plane, which leads to a reduction in brain activation by repetitive stimuli, but which operates, not peripherally, but in the zone between the cerebral cortex and the limbic and reticular activating systems. The following may be quoted as an example of habituation. If a note of a regular pitch is sounded close to a sleeping cat, the cat will wake up the first time. If the same note is now sounded at regular intervals the effect on the cat will gradually diminish. If, however, the pitch of the note is changed, its original waking effect will be restored, whereas a note of the original pitch will still fail to wake the cat. This experiment shows that identical stimuli lose their effect with repetition, provided that the stimulus is meaningless, i.e., has no significance in the life of the animal. This is the essential nature of habituation, the elimination of reactions to meaningless stimuli.

Habituation is a protecting filter

The mechanism of habituation may be compared to a filter which filters out stimuli that are meaningless in the life of the animal, allowing to pass only those that have a certain relevance.

The biological significance of habituation is the same as that of adaptation: the protection of the cerebral cortex (and thereby the entire organism) against being inundated with irrelevant alerting or alarm stimuli. Without habituation, the organism would need to maintain itself constantly in a state of maximum alertness.

It is an obvious assumption that the process of habituation also has a part to play in monotonous situations, when it nullifies the effect of irrelevant and repetitive events.

It appears from these considerations that adaptation and habituation are neurophysiological mechanisms which are to be taken as indicating the existence of monotonous conditions. *Situations in industry and in transport which give rise to the phenomena of adaptation and habituation certainly involve an increased risk of monotony and boredom.*

The neuro-physiology of boredom (summary)

The physiological aspects of boredom may therefore be summarized as follows. Situations which are characterized by a low level of stimulation, or by a regular repetition of identical stimuli, or just by making few mental or physical demands upon the operator, lead to a diminution in the flow of afferent sensory impulses, as well as to a lower level of stimulation of the conscious spheres of the brain. The consequent fall in activation level of the reticular and limbic systems reveals itself in a reduced reactivity of the entire organism. It is hardly possible to distinguish between fatigue and boredom on physiological grounds, since both of these states are characterized by a reduction in the level of cerebral activation. Nevertheless, differences do exist, as will be shown later, on the evidence of results from certain experimental work on boredom.

Medico-biological aspects of boredom

Until a few decades ago, the science of work physiology was mainly interested in finding out how to relieve the worker of excessive physical load. *Increasing mechanization and automation, as well as the tendency to divide up the work into as many simple operations as possible (Taylorism) has now led, in many occupations, to a new problem: insufficient demands on the physical and mental capacities.* Unused physical and mental capacities characterize a state which we call 'underload'.

Nearly all the organs of the human body have the important biological characteristic of being able to respond to stress by stepping up their performance. This is true not only of the muscles, heart and lungs, but also of the brain. Human development from childhood onwards is heavily dependent upon this ability to adapt to the stresses of life.

Conversely, if an organ is not exercised, it atrophies. A good example is the wasting of muscle which becomes distinctly noticeable only a few weeks after a fracture of a limb. Cessation of development, followed by decline, takes place on a mental level as well as a physical one. Thus it is known from experiments on animals that the brain becomes better developed, functionally, morphologically and biochemically, when the animal is subjected to various mental demands and stresses than when it is allowed to grow up in a quiet situation with few external stimuli.

From these considerations it is therefore evident that underload, such as a person experiences from monotonous, repetitive work, is basically unhealthy from a medico-biological standpoint.

The relationship between stress and biological reactions can be broadly summarized as follows:

1. *Underload leads to atrophy.*
2. *The right amount of load leads to healthy development.*
3. *Overload wears out the body.*

Boredom and adrenalin

An interesting contribution to a better understanding of the different aspects of monotonous work was made by several Swedish studies like those by Levi [166] and Frankenhäuser [72]. They analysed the catecholamine excretion in the urine and found that the most diverse physical and emotional stress situations led to a measurable increase in the adrenalin excreted in the urine, which was interpreted as a mobilization of the performance reserves of the body. One study by Frankenhäuser *et al.*, [73] is mentioned here because it is relevant to the problem of boredom. Their experiment on mental under- and overload yielded the following results:

1. *Overload*, created by a long-lasting serial reaction time test, produced an increased flow of adrenalin (about $9 \cdot 5$ ng/min).
2. *Moderate load*, in the form of reading a newspaper, gave only a small increase in adrenalin excretion (about 4 ng/min).
3. *Underload*, as a consequence of a uniform, repetitive operation also produced a higher flow of adrenalin, amounting to about $5 \cdot 7$ ng/min, and so falling between the levels of 'overload' and 'moderate load'.

Marianne Frankenhäuser concluded from this: 'The results show that adrenalin production is increased not only when acting under pressure, against the clock, and with a high inflow of information, but also in conditions that are monotonous, and lacking in stimulation. This shows that the physiological reaction is produced by the mental and emotional stress, rather than by the physical effort as such.'

Sawmill work and adrenalin

A field study by Johansson *et al.* [136] also produced interesting results: a group of sawmill workers, whose work was repetitive and at the same time responsible, secreted much more adrenalin than other groups of workers. They also exhibited a higher incidence of psychosomatic ailments and more absenteeism. The authors concluded that the combination of monotonous, repetitive work with a higher level of mental stress called for a continuous mobilisation of biochemical reserves, which, in the long term, adversely affected the general health of the workers.

13.3. Field studies and experiments

Laboratory studies have the advantage that they are conducted under controlled conditions, and therefore it is usually possible to distinguish clearly between 'cause' and 'effect'. On the other hand they have the drawback that the working conditions studied are usually simulated, and do not compare exactly with either industrial practice or everyday traffic. This reservation is particularly necessary when studying boredom. Even if it is possible to reproduce the physical working conditions very closely, the important psychological factors, e.g., the profit motive, or the role of social contacts, can be simulated only roughly in the laboratory. On the other hand it is usually possible to carry out field studies under everyday working conditions, although even so many factors are not under the investigator's control. Some results may be difficult to assess, or may be accepted only with reservations.

The following account concerns two field studies into the effects of monotonous work on the operators.

On the production line in factories

In 1960 Haider [101] studied boredom among 337 female workers in various factories, 207 of whom worked on a moving production line, while the other 130 worked individually. He assessed the subjective feelings of these workers by using a self-assessment card, with 12 pairs of contrasting states. Comparison of the two groups gave, on average, the following differences: the production line workers were more 'tense', 'bored' and 'dreamy' than those who worked at their own speed. Haider writes of these: 'Finally we may conclude that when the occupation is a long succession of simple, repetitive acts, we may expect to see 'satiation phenomena', with increasing tension, restlessness, lack of incentive, and a declining performance of the boring work'. Indeed, the percentage of discontented and tense workers on the production line was as high as 20−25%.

Checking bottles

Saito *et al.* [227] carried out a similar investigation in the food industry. The operatives studied were engaged in the visual control of bottles and their contents which moved so quickly that the job was rated as arduous, even though monotonous and repetitive. Even after a short time at work the number of rejected bottles became distinctly lower, a fact which the authors considered must indicate diminishing alertness. Concurrently a distinct fall in the flicker-fusion frequency could be recorded, accompanied by such subjective symptoms as increasing fatigue, sleepiness, headache, and a sense of 'time dragging'. The fall in flicker-fusion frequency was greatest at times when the operative was observed to talk less to the others and tended to doze, or even to fall asleep.

The progression of all effects was as follows:

First hour: no change.
2nd—4th hours: marked impairments.
Final hour before lunch-break: operatives both felt and
performed better.

Experiments on boredom

The following are a few selections from the very large number
of important experiments that have been carried out to
investigate boredom and its effects on people at work. Firstly
we must mention the many vigilance studies already discussed in Chapter 10. In practice nearly all these involved a
monotonous task which nevertheless called for a constant
state of alertness, and they have overwhelmingly shown that
prolonged concentration on a monotonous task results in a
steady decline in alertness. In the same chapter it was also
pointed out that vigilance is dependent upon the functional
state of the brain, the level of cerebral activation.

In order to induce a state of boredom Hashimoto [104] used
simulated driving tests. One test series was easy, with the
second one much more difficult. The easy condition produced
a distinct lowering of the flicker-fusion frequency and an
increase in alpha rhythms on the electroencephalogram, as a
result of boredom. Under the more difficult condition boredom
was less evident.

Experiments with simulated repetitive tasks

Whereas the experiments described above were predominantly concerned with simulated traffic conditions, Martin
and Weber [179] and Baschera and Grandjean [11] elected to
produce their state of boredom by means of a very uniform
and repetitive task. Over a period of several hours, their test
subjects were required to pick up nails singly, count them, and
place a specified number into a series of envelopes. This task
fulfils many of the conditions conducive to boredom: it is
extremely repetitive, is undemanding, and does not require
much alertness, yet it does not leave the mind entirely free for
daydreaming.

In all these nail-counting experiments the work caused a
lowering of the flicker-fusion frequency. Simultaneous checks,
either with a questionnaire or a bipolar self-assessment chart,
showed changes in the subjective well-being — either less
efficient and less willing manipulation of the nails, or a
displacement of the bipolar charts in the direction of increased
sleepiness, fatigue, inattention and boredom.

Figure 125 reproduces some of Martin and Weber's work
[179], in which the results of the nail-counting experiment
were compared with a study of more stimulating conditions,
which required the research subjects to carry out a variety of
psychomotor tests, in the intervals of which they listened to
music of their own choice.

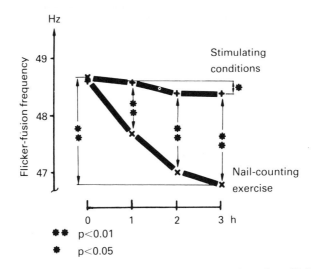

Figure 125. Average values of the subjective flicker-fusion frequency of the eyes of 25 test subjects in a nail-counting exercise, as compared with a stimulating situation.
After Martin and Weber [179].

As the diagram shows, the monotonous counting of nails quickly provoked a distinct fall in the flicker-fusion frequency, which reached an average value of 1·7 Hz after 3 hours. In contrast, the more stimulating conditions of the second experiment produced only an insignificant fall in flicker-fusion frequency in the same length of time. Similar results were obtained from questioning the test subjects about how they felt. The nail-counting experiment produced an increased sense of strain (through loss of manual dexterity) and a loss of motivation (disinclination for the task): whereas almost all the effects of the more stimulating conditions were beneficial.

Research therefore shows that *a monotonous, repetitive job can quickly lead to boredom, as shown by a fall in flicker-fusion frequency and changes in subjective feelings.*

Both in driving tests [104] and in field studies among drivers of motor vehicles [102, 206] boredom could be avoided or at least postponed either by stimulating conditions, or by making the task more difficult. This state of affairs leads one to ask whether repetitive jobs could be made less boring simply by making them more difficult. To test this we extended the nail-counting test by introducing a moderately difficult and a very difficult mental task [11]. The test subjects were required to perform the following three tasks, each lasting three hours:

Repetitive tasks with varying mental demands

1. The nail-counting test as described above: *low mental demand.*
2. Select nails of different colours and with different ring markings in such a way as to fill an envelope with the

correct number, colour and ring marking. We rated this
exercise as making only a *moderate mental demand*.

3. Select nine nails, with a prescribed combination of colours
    and markings, and stick them into a board in an ascending
    order of rank according to their ring markings. This was
    the exercise that made *heavy mental demands*.

Figure 126 shows the average fall in subjective flicker-fusion
frequency of the 18 research subjects who were tested under
the three sets of conditions.

The outstanding result of these three series of tests is the
distinct relation between the flicker-fusion frequency and
mental stress: the low stress level is accompanied by the
greatest fall in flicker-fusion frequency; the high level of
mental stress equally provokes a fall in this frequency, though
it is somewhat smaller than before. On the other hand, the
moderate mental stress leaves the flicker-fusion frequency
almost unaltered. We can draw a theoretical U-curve through
the tops of the three columns in the block diagram, which
suggests that both understress and a high level of overstress
cause a reduction in the level of cerebral activity. Between
these two extremes lies a zone of moderate mental stress
which has no adverse effect on the functional state of the CNS.
The analysis of the simultaneously filled in questionnaires
related to subjective feelings revealed some interesting
results. Several feelings, e.g., weariness and sleepiness, show a
parallel with flicker-fusion frequency. They were increased

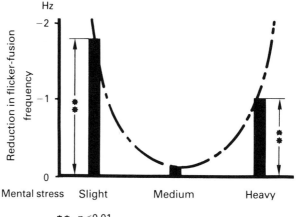

** p<0.01

**Figure 126. Average reduction in subjective flicker-fusion frequency
of the eyes during repetitive tasks which involve various levels of
mental loads.**
*The vertical columns express the difference in the measurements
before and after the 3·5-hour experiments, involving 18 test subjects.
The broken line is a hypothetical curve expressing the probable
course of flicker-fusion frequency in relation to mental load. After
Baschera and Grandjean* [18].

both for low and for high levels of mental stress. In contrast, 'boredom' was predominantly associated with low level of stress. This shows that it is obviously possible to differentiate between the effects of under- and overstress by asking the appropriate questions. It is doubtful, however, whether it is yet possible to make this distinction by physiological methods.

These results are in accordance with those of Frankenhäuser *et al.* [73] mentioned above, showing an increased adrenalin secretion not only under work pressure but also in monotonous conditions.

14. Job design in monotonous tasks

14.1. The fragmented work organization

Taylorism and boredom

For several years, more and more critical objections have been raised by social scientists to the Tayloristic principle of splitting a job into a large number of identical tasks which are repeated indefinitely. Workplaces organized on this principle are characterized by low cycles per piece and few demands on the operative. *The result of such a fragmentation of the job is that individual freedom of action is severely curtailed, mental and physical abilities lie fallow and the potential of the worker is wasted.*

Several surveys favour the hypothesis that there is a link between the quality of one's working life and that of one's life in general. Comments on these social and ethical aspects of job satisfaction can be found in the work of Friedmann [75] and in the *Report of the International Conference on Enhancing the Quality of Working Life* [211].

Some escape into daydreams and like it

Some studies also revealed that there are always individuals who rate their repetitive and monotonous work 'interesting'! This confirms the experience of many factories that a certain proportion of the workers enjoy their monotonous, repetitive jobs and do not want one that is more varied or more challenging. It seems that some individuals are able to escape with their thoughts into a world of daydreams and that they appreciate working conditions which permit them to do so. On the other hand, personnel managers report that it is becoming increasingly difficult to find workers to do the monotonous and repetitive jobs.

It can be concluded that different attitudes really exist in practice. For an indefinable proportion of workers, male or female, working on a production line must actually be more relaxing than free assembly, since it allows them to express their personalities better by conversation and daydreaming. For another proportion of workers, however, continuous work on a production line seems meaningless because it does not provide them with opportunities to develop their personalities by exercising their brain power at work.

Whatever the individual preference may be, all social scientists and occupational psychologists agree that a job that takes account of a person's potential and inclinations will be carried out with interest, satisfaction and good motivation. Conversely, it is obvious that an undemanding job, which does not develop the potential of the worker will prove to be boring and lacking in motivation. At the other extreme, a job that requires more from the worker than he or she is capable of will be overtaxing. Hence the obvious requirement is that work should be so planned that it is matched to the capabilities of the operatives, without asking either too little or too much of them.

This basic recommendation embodies the idea that the efficient performance of a complex or difficult task is at its optimum in the range between being under- and overdemanding. This concept appears in Figure 127 as an inverted U-shape.

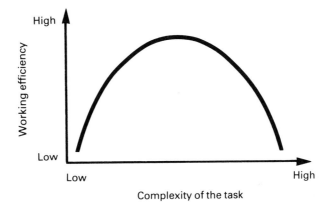

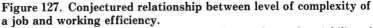

Figure 127. Conjectured relationship between level of complexity of a job and working efficiency.
The level of complexity is deduced from the number and variability of the operations involved.

Similarly Blum and Naylor [20] concluded that the level of frustration in relation to the level of complexity of the task was represented by a U-shaped curve, and that frustration was least when the demands of the job were most closely matched to the capabilities of the worker.

Summary of the consequences of extreme fragmentation of work

Table 23 summarizes the most important critical objections, culled from various branches of science, against extreme Tayloristic fragmentation of human tasks in monotonous, repetitive work.

Table 23. **Monotonous, repetitive jobs from the viewpoint of various sciences.**

As seen by	Probable consequences
Doctor	Atrophy of mental and physical powers
Work physiologist	Boredom; risk of errors and accidents
Occupational psychologist	Increasing discontent with the job
Social scientist	Human potentialities not fully realized
Industrial engineer	Increased absenteeism; increasing difficulty in finding personnel to do the job

14.2. Principles of job design

The various consequences that may follow from repetitive work have led in recent years to the development of different ways of organizing and restructuring assembly work and similar serial jobs.

Aims of restructuring job design

The main objective of these efforts is to give the operator more freedom of action in the following two directions:

1. *Reduction of boredom*, with its concomitant feelings of fatigue and satiation.
2. *Making the work more worthwhile* by providing a meaningful job, and one which allows the operator to develop his or her full potential.

Basic to both these improvements is the assumption, already indicated above, that there will be a reduction in absenteeism, turnover of the work force and social stress, and that the new conditions will attract more workers. Hence in the long run higher productivity should result.

These desirable new forms of work organization involve various improvements, ranging from simple variety of work, through various ways of broadening the scope of the job, to job enrichment by giving the worker more information, more responsibility, more participation in decision-making and more control of the work process.

Increase of variety of work

A first step to improve repetitive working conditions is the attempt to increase the variety of work. This is a scheme by which each individual worker is entrusted with different jobs at different workplaces, which he or she carries out in rotation. Such attempts have been made in industry: for instance, the rotation of workers round a work bench with frequent changes of assembling operations. The best solutions were judged to be those which at the same time introduced a certain group autonomy, allowing the workers to control the making of their own product.

An example may be taken from the assembly of electronic calculators. The complete assembly is deployed round a work bench. There are eight places, but only six operatives, so that there are always two vacant seats. The resulting accumulation of components forces the female operatives to change their seats continually. It is an essential feature of this system that each operative must be trained to work at any of the eight places, and so the complete assembly is a function of the entire group.

However, one point must be emphasized: if variety of work merely means moving to and fro between jobs that are equally monotonous or repetitive, the risk of boredom may be slightly reduced, but the desirable matching of the difficulty of the job to the capabilities of the worker is not being achieved. Adding yet another monotonous, repetitive job is not going to lead to job enrichment!

Broadening and enriching the job

For these reasons a special importance attaches to types of organization which strive to enrich the work by broadening its scope, thus helping to develop the personalities and self-realization of the employees. In such organizations the tasks are so planned that a worker moves to a succession of different jobs, each of which makes different demands on his or her abilities. Additional responsibilities such as quality control, or the installation and maintenance of machinery, make an important contribution to the enrichment of the job. Many examples of such broadening and enrichment of employment can be found in the literature [211].

An actual observation in a factory making electrical apparatus may serve as an example. A certain piece of equipment was originally assembled on an assembly line by six successive operations performed by six workers. In the new plan, one and the same worker carried out all six operations, being solely responsible for the quality of the entire assembly.

Autonomous working groups

A further step towards more worker participation was the organization of autonomous working groups. The workers employed for each production unit were organized in a group and the planning and organization of the work as well as the control of the end-product were delegated to them. The autonomous working group, therefore, has planning and control functions.

Here is a Dutch example [109]. A firm making television sets introduced more attractive forms of work organization, by stages, as it became a problem to find workers. At first the sets were assembled on a long assembly line, and the 120 workers pushed their set along once a minute, at a signal. The first stage of improvement was to have a new assembly line with 104 working places, divided into five groups. At the end of each group there was a barrier and a quality check. It was found that with this arrangement assembly times were shortened, and quality improved. Next the experiment was tried of allowing more time before the sets moved on, as well as a

certain amount of job-switching. The final step was to introduce *autonomous working groups*. The groups were reduced in
size to seven people, and the range of each person's tasks
broadened, so that intervals between moves rose from 4 to 20
minutes. The working groups undertook many responsibilities
which previously had fallen upon the foreman or overseer. The
groups performed their own quality control, and were self-
administering. A certain amount of interchange within the
group made them more flexible.

The authors declare that the first results include:

1. More positive approach to the work.
2. Greater co-operation among members of a group.
3. Creation of a certain critical attitude to production levels.
4. The groups were a nuisance to the management in one
   sense, since autonomous groups acted as a controlling
   force on the management.
5. Less absenteeism.
6. Less waiting time.
7. Fewer 'passengers'.
8. More working areas and new machines.
9. More consultation.
10. Higher wages.

Taken as a whole, the economies predominated and the
groups produced a television set more cheaply than did their
predecessors.

All efforts to broaden and enrich people's work must be
regarded as experiments that are by no means completed, nor
can their results as yet be fully evaluated. There are reports on
successful attempts, but there are also cases where such
projects had to be given up because of resistance on the part of
employees, unions or managers. In fact, the search for new
ways of restructuring monotonous, repetitive and meaningless work is still going on.

Social contacts   This catalogue of organizational improvements may be concluded by emphasizing the importance of social contacts at the
workplace. *The opportunity to talk to one's fellow workers is
an effective way of avoiding boredom.*

Conversely, social isolation brings monotony and increases
the tendency to become bored with the work.

Sitting one behind the other along a straight assembly line
is bad: it is much better if the line follows a semi-circle, or is
sinuous. Any arrangement is good so long as it brings several
workers within conversational distance of each other.

Other ways of reducing the incidence of boredom include:

1. More frequent short stops.
2. Opportunity to move about during these stops.
3. A stimulating layout of the surroundings, making use of
   light, colour and music.

Job design for supervisory work

The same principles apply to supervisory work and to driving a vehicle. An essentially dull situation, lacking in stimuli, must be enlivened just enough so that it is neither soporific on the one hand, nor overstressed on the other.

In both types of job the problem is often to be able to recognize a critical situation at the right moment. The necessity to remain alert, which is often associated with a high level of responsibility, results in both fatigue and boredom, and increases the risk of mistakes and accidents. Hence preventive measures concentrate on making sure that relevant items of information are correctly assimilated.

To summarize, the following arrangements can be recommended for supervisory jobs at control panels, machines, projection screens and similar places of work:

1. Alarm signals and similar safety limits must be clear and decisive. A combination of a light signal and a noise (buzzer, gong or siren) is particularly effective.
2. If a series of signals must be noticed without missing any of them, there should be between 100 and 300 of them per hour.
3. The operator must be fresh, and must avoid getting tired beforehand. Night-shift workers are particularly liable to boredom until they have become night-adapted.
4. The surroundings should be brightly lit. Music is helpful in the right circumstances. Room temperatures should vary only within comfortable limits.
5. A change of work must be arranged as soon as boredom causes dangerous lapses of alertness. In extreme situations it may be necessary to consider hourly changes.
6. Short pauses — as far as these are possible — will help to avert boredom and improve alertness.
7. In certain particularly critical situations it may be necessary to employ two people to keep a look-out together, e.g., in the driver's cab of a locomotive.

And what about VDT jobs?

As shown above, attempts have been made in industry to avoid adverse effects of repetitive, monotonous jobs by means of alternative ways of organizing production. Gunn Johansson [134] writes: "Unless counter-measures are taken also in the white-collar sector *there is a risk that the new technology* [of VDTs] *will create highly repetitive tasks which will require little skill, allow little social interaction, and generate the type of negative consequences associated with mechanized mass production. Data-entry work is a case in point"* (emphasis added).

Many authors agree with these words of warning and point out that some clerical jobs had little content before the era of VDTs, and that VDT jobs are often fragmented and simplified versions of traditional clerical activities. This is certainly true for some very simplified data-entry or data-acquisition tasks, whereas many other VDT jobs are characterized by a high degree of complexity, judged interesting and challenging by operators.

Job redesign necessary for data entry or data acquisition

Attempts to improve job design are therefore justified mainly for the highly repetitive and monotonous data-entry and data acquisition jobs.

No projects or results of restructuring fragmented and repetitive VDT jobs with the aim of improving job design have been published yet. For that reason claims in connection with job design are mainly based on general considerations of the relationships between working conditions and job satisfaction.

Broadening data-entry tasks is certainly rather difficult. Some banks have improved the situation by creating 'mixed jobs', alternating pure data entry with payment transfers and other more demanding tasks. Furthermore, rest rooms have been provided, where operators can take their breaks together. Other banks have engaged only part-time employees for pure data-entry jobs, which, however, has been less successful than 'mixing activities'.

Control of work process

Lack of job control seems to be an important social stressor. Job control means participation and increased decision making. Performance feedback is a vital part of the worker's control of the work process. It is recommended that an operator should receive feedback direct via his or her own VDT screen. Such direct performance feedback seems to enhance job satisfaction. If, on the other hand, this information is forwarded to the supervisor, the employee will experience this as lack of control of his or her work; the supervisor might use the information to control the worker, thus creating more social tensions.

VDT work should be meaningful

The meaning or content of work may be low in some VDT tasks. As work is fragmented, it is also simplified. Thus workers fail to identify with the job and lose interest in the product of their work. If fragmentation cannot be avoided, it is important that employees at least understand their contribution to the end product. They should feel that their contribution is important, which will heighten their satisfaction and self-esteem.

Social contacts should be granted

One of the obvious drawbacks of VDT work is the low chance for social contacts, particularly with colleagues. Many VDT jobs lead to an isolation of individual operators, much more than traditional clerical activities. Therefore it is advisable to enhance and encourage social interaction during non-task periods; this is an argument in favour of work breaks with facilities in the neighbourhood.

A careful introduction policy

A careful and well-planned introduction of VDTs is an important measure to prevent hostility towards office automation. A good transition policy should include proper information and clear instructions, adapted to the worker's capabilities. It is certainly insufficient to leave the employee alone with a manual that explains how the system works. Classroom teaching, followed by practical application, should be given by a well-qualified expert able to coach the trainees. A good training programme will increase acceptance and reduce psychological fears, since well-trained operators will come to consider themselves an important investment.

Some restrictions must be emphasized

Some restrictions must be made, though, when discussing the above-mentioned principles of job design for VDT work. It must not be overlooked that the main problems of job design refer to some repetitive, monotonous and meaningless data-entry or data acquisition jobs. The proportion of these among the total number of VDT activities is not known, but a figure of between 15 and 25% is a reasonable assumption. Another restriction concerns the above-mentioned fact that not every employee dislikes repetitive jobs. Salvendy [228] reports that 10% of the labour force in America do not like work of any type; the remaining workers are evenly split between those who prefer to work in enriched jobs and are more satisfied and productive in them and those who prefer to do simplified jobs in which they are more satisfied and productive.

15. *Working hours and eating habits*

15.1. Daily and weekly working time

Daily working time

Many studies have shown that changes in the length of the working day often result in higher or lower output. Thus at one factory it was found that shortening the working day from $8\frac{3}{4}$ to 8 hours raised the output by between 3 and 10%, those whose work was predominantly manual showing a bigger increase than machine operators.

Figure 128 sets out the results of an old British survey [263]. These results confirm the general experience that shortening the working day results in a higher hourly output; the work is finished off more quickly, with fewer voluntary rest pauses. This change in working rhythm generally takes place within a few days, although occasionally it may be several months before the effect can be seen.

Conversely, making the working day longer causes the tempo of the work to slow down and the hourly output to fall. The relationship between a day's work and a day's total output has been worked out, to a first approximation, by Lehmann [162] (see Figure 129). From this it can be established that as a rule the total daily output does not increase in proportion to the daily working time (curve A, Figure 129); most often it behaves like curves B and C. In many cases it has even been found that increasing the daily hours beyond 10 results in a fall in total output, because the slowing down of the work tempo resulting from fatigue more than offsets the longer hours.

A longer working day or overtime working are common in wartime and in boom periods, but the results are often disappointing. Because of the relationship between hours and output just mentioned, productivity does not increase as much as is desired and expected, and it may even fall.

All these observations point to the conclusion that the worker tends to maintain a particular daily output, and if the working day is varied, will to some extent adjust his or her working rhythm to compensate. This is only true, however, for work that is not linked to the speed of a machine. Workers on a

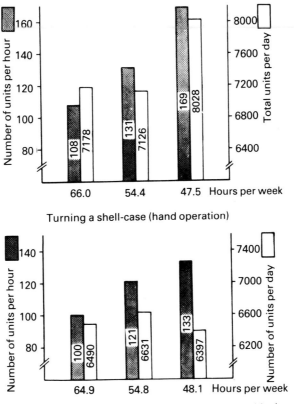

Turning a shell-case (hand operation)

Milling a screw-thread (mainly operating the machine)

Figure 128. Working time and output.
Top: for predominantly hand work, shorter hours (either weekly or daily) bring about an increase in both hourly and overall daily output. Bottom: for a job that is mostly operating a machine, shorter hours, while they increase hourly production, have little effect on the overall daily output.

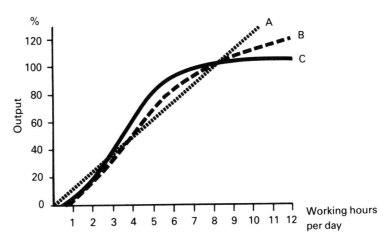

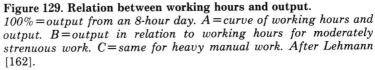

Figure 129. Relation between working hours and output.
100% = output from an 8-hour day. A = curve of working hours and output. B = output in relation to working hours for moderately strenuous work. C = same for heavy manual work. After Lehmann [162].

conveyor belt, and others who must fit their work into the rhythm of a machine, cannot do much to compensate for changes in working times. This can be seen from Figure 129.

Of course, the extent to which working speed is varied to compensate for variations in working hours is also affected by rates of pay and other changes in motivation.

Effects on sickness rates

Many observations have provided evidence to show that excessive overtime not only reduces the output per hour, but is also accompanied by a characteristic increase in absences for sickness and accidents. A working time of eight hours per day, which makes the operator moderately, but not seriously fatigued, cannot be increased to nine hours or more without ill-effects. These include a perceptible reduction in the rate of working, and a significant increase in nervous symptoms of fatigue, often resulting in more illnesses and accidents.

Our physiological knowledge and present-day experience point to the conclusion that a working day of eight hours cannot be exceeded without detriment, particularly if the work is heavy. Modern firms, organized according to the principles of industrial science, usually arrange most of their work sensibly, whether the demands on their workers are heavy or only moderate. An extended working day is tolerable in jobs where the nature of the work provides plenty of rest pauses.

Historical review of weekly work-ing time

The duties of an employee are usually agreed as so many hours per week, which have become markedly shorter during the last hundred years.

In Switzerland, the first federal Factory Act was accepted in a plebiscite in 1877, and stipulated 65 hours per week (11 each weekday and 10 on Saturday). An amendment in 1914 short-ened this to a 48-hour week.

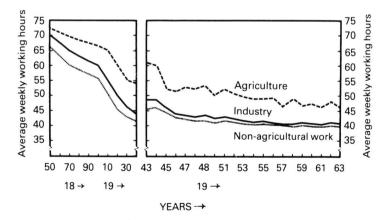

Figure 130. The trend of weekly working hours in the USA from 1850 to 1963
After Northrup [205].

Figure 130 shows the historical development of weekly working hours in the USA.

It is evident from this diagram that the working week in the USA has tended to shorten almost continuously since 1850. Nowadays the 40-hour week or less is common, not only in the USA but in most industrial countries.

The five-day week

Nowadays the five-day week is common everywhere, and it presents little problem to combine this with a 40-hour week. Nevertheless the experience of introducing the five-day week is of some general interest. Occasionally factories found that the changeover from a six- to a five-day week led to less absenteeism. Experience shows that the workforce in general, and especially women, prefer the five-day week, mainly on social grounds. These social factors, combined with increased opportunities for rest and relaxation, are mostly responsible for the reduced absenteeism.

The four-day week

A four-day week has come under discussion in recent years. About 600 US firms and recently some German and French ones, are said to have had favourable experience of this [178, 223]. The total hours worked in four days are usually 36 in the USA and 40 in the European firms. Rosenkranz [223] thinks that the trend is towards a four-day week of 36 hours, but that industrial conditions are not yet favourable to the idea. Advantages put forward are three days free at the weekend, and the possibility of increasing employment by taking on more workers.

Today's worldwide unemployment is certainly an important argument in favour of such a four-day week.

A reduction in the weekly working days to four or even three will, in most cases, be accompanied by a considerable increase in daily working hours. Some medical doctors as well as ergonomists consider such changes to be damaging to health. This criticism is certainly valid. It must be possible for a person to recuperate within each 24-hour period, not to go on exhausting himself (or herself) for four days and hoping to recover over the following three rest days. We now know that working days of nine or ten hours lead to excessive fatigue and increased absenteeism through sickness. *The four-day, 40-hour week must therefore be rejected on medical and physiological grounds.* Moreover, there are organizational aspects to be considered. Compressing the working week into four days would throw up very severe problems, one of them that the recruiting of more workers would not suit every firm!

Maric [178] reported on the US experience, which is said to have been very mixed. The results of a public opinion survey in Germany [4] in 1971 yielded the following figures of preference:

Four-day week × 10 hours:	46%
Five-day week × 8 hours:	47%
Uncertain:	7%

It is very doubtful whether a similar result could be expected today, because in the meantime changes in global trading patterns have led to more short-time working and more unemployment, with consequent changes in working hours.

Working times in the future

Analysis of sociological trends has led many students of the future to believe that the working week will shorten still further. Thus, for example, the American forecaster Kahn [138] makes the prognosis that by the year 2000, weekly working hours, not only in the USA but also in Europe, will have been reduced to 35 or even 30, and annual holidays extended to 2−3 months. The main arguments for this belief are as follows.

The employees want to profit from the expected boom not only in pay, but also by having more leisure. A higher income makes sense only if the money can be used to enhance the quality of life. The boom in tourism and the so-called leisure industries in the last 20 years reinforce this argument. Less certain is the view that as modern industry becomes more 'rationalized' the work will be less satisfying and less stimulating, so that there will be more need to give a meaning to life in one's leisure hours.

Most of these prophecies were made before 1970, since when we have become more acutely aware that industry, the consumption of energy, and the destruction of natural resources cannot go on increasing at such a rate. Hence one of the most important arguments for shorter hours — that they would increase production and speed up industrial growth — is being called into question.

Flexible working hours

Flexible working hours are a recent form of work organization which has found many adherents during the last 20 years or so. It is characterized by a fixed working period, called the block time or core time, and a flexible period at each end of this, the length of which is at the discretion of the individual worker. All the employees must be present during the core time, so obviously a certain number of hours must be worked per month, as a minimum. The weekly or monthly totals are recorded by various methods such as stamped cards, click-counters or punch cards.

A good review of experiences in various countries is to be found in the publication of the International Labour Office by Maric [178]. On the whole it can be said that this arrangement is popular with the employees.

A valid objection is the fact that not all kinds of work are suitable for flexible hours, and that in every factory there is at least one group of employees who must be excluded (for example, workers at the information desk, telephonists, certain servicemen, switchboard operators, etc.) According to Maric [178] other groups for whom flexible hours are inappropriate are production-line operatives working in series and other assembly workers organized in groups.

Looking over the very extensive literature on flexible working hours, one gets the distinct impression that the advantages greatly exceed the drawbacks, and that both on industrial and social grounds this new way of regulating working hours is a distinct step forward.

15.2. Rest pauses

Biological importance

Every function of the human body can be seen as a rhythmic balance between energy consumption and energy replacement, or, more simply, between work and rest. This dual process is an integral part of the operation of muscles, of the heart, and, if we take all the biological functions into account, of the organism as a whole. Rest pauses are therefore indispensable as a physiological requirement if performance and efficiency are to be maintained.

Military commanders have always known that a marching column should be halted once an hour, because the time lost will be more than compensated for by a better performance by the men at the end of the march. Rest pauses are essential, not only during manual work, but equally during work that taxes the nervous system, whether by requiring manual dexterity or by the need to monitor a great many incoming sensory signals.

Different kinds of rest pause

Work studies have shown that people at work take rest pauses of various kinds and under varying circumstances. Four types can be distinguished:

1. Spontaneous pauses.
2. Disguised pauses (i.e., switching to routine work for a time).
3. Pauses arising from the nature of the work.
4. Prescribed pauses, laid down by the management.

Spontaneous pauses are the obvious pauses for the rest that the workers take on their own initiative. These are usually not very long, but may be frequent if the job is strenuous.

Disguised pauses are times when the worker occupies himself or herself with some easier, routine task in order to relax from concentrating on the main job. Most jobs offer many opportunities for such disguised pauses, examples being the following: cleaning some part of the machine, tidying the work bench, sitting down more comfortably, or even leaving the workplace on the pretext of consulting a workmate or the foreman. Such disguised pauses are justified from a physiological point of view, since nobody can do either manual or mental work continuously, without interruption.

Work-conditioned pauses are all those interruptions that arise either from the operation of the machine, or the organization of the work; for example, waiting for the machine to complete a phase of its operation, for a tool to cool down, for a piece of equipment to warm up, for a component, or for a machine or a tool to be repaired. Waiting periods are especially common in the service industries, e.g., waiting for customers or orders. On a conveyor belt the length of work-conditioned pauses depends on the speed and dexterity of the operative. The faster he or she works, the longer he or she has to wait for the next piece to come along. Since speed and dexterity decline with age, the younger workers have long pauses, whereas the older operatives often have to work almost continuously to keep up. Hence on conveyor belts, the older operatives, as well as the less skilled often have to work hastily, and may be overstressed.

Prescribed pauses are breaks in the work that are laid down by the management, for example, the midday break, and other pauses for snacks (e.g., coffee break).

Interrelationship between pauses

The four types of rest pause are to some extent interrelated, as was shown by Graf [81] with the help of a time study. In particular, this showed that the introduction of a few prescribed pauses led to fewer spontaneous and disguised pauses being taken. Figure 131 shows the results of a time study of a female operative in the electrical industry, who had to perform a highly skilled job on piece rate.

In a similar investigation Graf [81] showed that during tiring skilled work standing up (pressing fuse-holders out of china clay) both disguised and spontaneous pauses increased progressively during the course of an eight-hour study. There were about three times as many pauses during the last three hours as during the first five, showing that the need for breaks in the work increased as the operator became more fatigued. To meet this need there was an increase in both disguised and spontaneous rest pauses.

In general, it can be said that all the different types of rest pauses (disguised, spontaneous, prescribed and work-conditioned) should amount to 15% of the working time. Often a ratio of 20—30% is allowed, and this is certainly necessary in some jobs.

Rest pauses and output

Many investigations into the effect of rest pauses on production have been recorded in the literature, and in general their results agree with those of research into working time and production. Introducing rest pauses actually speeds up the work, and this compensates for the time lost during prescribed

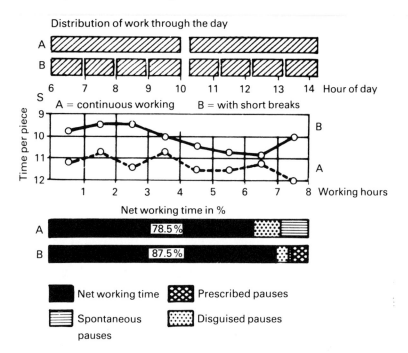

Figure 131. **Effects of short breaks (B) on the net working time (black block), on secondary tasks (disguised pauses) and on spontaneous pauses.**
After Graf [81].

pauses, as well as leading to fewer disguised and spontaneous pauses.

The hourly output of fatiguing work usually declines towards the end of the morning shift, and even more towards evening, as the rate of working slows down. Various studies have shown that if prescribed pauses are introduced the appearance of fatigue symptoms is postponed and the loss of production through fatigue is less.

Even though not every investigation has been conducted on strictly scientific lines, they have indicated that *on the whole rest pauses tend to increase output rather than to decrease it. Ergonomics attributes these effects to the avoidance of excessive fatigue, or to the periodic relief of fatigue symptoms by an interval of relaxation.*

Rest pauses in heavy work...

For heavy work obligatory pauses should be laid down, evenly distributed throughout the eight working hours of the shift. If the pauses are only optional, workers tend to work continuously and save up all the permitted rest time until the end, so as to be able to leave work earlier. This leads to overstress, particularly among older workers. The rest pauses must be arranged in such a way that the total energy expenditure of 20 000 kJ per working day is not exceeded.

...and in moderately heavy work

For all other jobs in manufacturing industries, in offices and in administration the current recommendation is a rest pause of 10−15 minutes in the morning, and often a similar interval during the afternoon. These pauses serve the following purposes:

1. *Preventing fatigue.*
2. *Allowing opportunities for refreshment.*
3. *Allowing time for social contacts.*

It is unthinkable not to have some rest pauses, which the workers value as much for social as for medical reasons.

Problem of time-linked jobs

Time-linked work on an assembly line poses a special problem. Many studies in the laboratory, as well as in factories [91], have shown that pauses of 3−5 minutes every hour reduce fatigue and improve concentration. They are especially necessary when the job is repetitive, has to be completed in a given time, and calls for constant alertness.

Under training

Rest pauses have a dramatic effect on the learning of a skilled operation, as we have already seen in Chapter 8. If training is interrupted frequently for short periods of relaxation, a new skill is acquired much more quickly than if training is continuous. Rohmert *et al.* [221] proved this in their monograph from much experience and many experiments.

Rest pauses during training do more than just prevent fatigue: during rest pauses a trainee will look ahead and understand the process, so that it becomes easier to acquire the automatic skills required. So where a skilled operation is concerned, the rest pauses provide additional periods for mental training.

This aspect should be considered when dealing with the training of apprentices, who benefit from frequent rest pauses.

Recommendations

To summarize, the following arrangements for rest periods are recommended:

1. *Where heavy work is concerned, or working in great heat, rest pauses should be such that the expected maximum hourly demands of the job should not be exceeded (see Chapter 6).*
2. *For jobs demanding moderate physical or mental effort, there should be a break of 10−15 minutes, both morning and afternoon.*
3. *A job making heavy mental demands, especially if it is timed work, or involves very little waiting time, should have, in addition to the morning and afternoon breaks, one or two shorter pauses of 3−5 minutes, both before and after midday.*

4. *When learning a skill, or serving an apprenticeship, many pauses should be the rule, varying in frequency and duration to suit the difficulty of the job to be learned.*

Time schedules for VDT jobs

The different opinions about the impact of working hours on the adverse effects of VDT work are controversial: some observed an increase of complaints with increasing working time at a VDT, others could not confirm such a relationship. Some unions and a few scientists have claimed that the number of working hours each day spent at VDTs needs to be reduced. A principle objection must be made here: *neither change nor reduction in working time should be considered until the display, workstation and environment fulfil the main ergonomic design recommendations.* It would be nonsense to reduce the working time because of a badly designed workplace. There are good reasons, at present, to believe that the work at ergonomically well-designed VDT workstations is not more strenuous than other office jobs.

Some unions claim hourly rest pauses for VDT operators. Such proposals could be considered for very repetitive and speedy data-entry jobs. But, as a rule, most VDT jobs are certainly comparable to the usual office jobs, for which two rest pauses, one in the morning and one in the afternoon shift, are recommended.

15.3. Nutrition at work

Comparison with a motor car

A car requires three essentials for smooth running:

1. *Petrol as a source of energy.*
2. *Lubricating oils and greases to protect the moving parts.*
3. *Cooling water.*

Similarly the 'human engine' requires:

1. *Foodstuffs (sugar, protein, fat) as a source of energy.*
2. *Protective materials (vitamins, mineral salts, iron, iodine, unsaturated fatty acids, etc.) as 'lubricants'.*
3. *Liquids for cooling purposes.*

Figure 70 demonstrates how chemical energy in the form of nutrients is taken in by the body and converted into heat and mechanical energy. This, again, is a comparable process to that which takes place in the engine of a car. A car can be driven only as long as the petrol lasts; a human being can go on working only as long as the food provides him or her with chemical energy. The more manual work done, the greater the demand for energy, which can be met only by increasing the intake of food.

The energy content of foodstuffs can be measured, and is expressed in kilojoules (kJ). (The item kilocalorie has also been commonly used in the past; 1 kcal ≈ 4.2 kJ.) The same unit is used for the energy consumption of the human body, which increases the more physically active is one's occupation. The average daily requirement of energy for men and for women in various occupations is summarized in Table 12 (p. 84).

In recent years the overall pattern of the working population has changed in regard to physical labour. In every industrial country the proportion of workers with sedentary jobs has greatly increased, until we can expect these to account for about 70% of employed persons. Conversely, the proportion of manual workers has fallen. A German example is given in Table 24 after Wirths [274], and this can be extrapolated into the future.

In general terms present-day adults can be divided into two categories according to their energy requirement in their occupation:

1. Sedentary workers and all female operatives: energy requirement 8400−12 500 kJ per day.
2. Heavy workers whose average daily requirement is 12 500−17 000 kJ (ignoring the few whose exceptionally severe work calls for 17 000−21 000 kJ per day).

Table 25 summarizes a proposal for distributing the requirement among the five recommended meal times.

Table 24. **Distribution of employed persons, including housewives, in FR Germany, in percentages.**
Figures from the states of Germany.

Severity of occupation	1882[*]	1925[*]	1950	1975
Light, sedentary work	21	24	58	70
Moderately heavy work	39	39	21	23
Heavy to severe work	40	37	21	7
Work force in millions	16·9	32·0	32·2	39·8

Table 25. **Distribution of daily food intake, in kJ.**

Meals	Sedentary workers and women	Manual workers
Breakfast	1200− 1700	2500− 2900
Morning break	100− 200	600− 1000
Midday meal	3300− 3700	3700− 4200
Afternoon break	100− 200	600− 1000
Evening meal	5200− 5900	5900− 7000
Total	9900−11700	13300−16100

Table 26. Energy equivalents of some important foodstuffs.
The quantity of each that must be eaten to give 420 kJ of energy is:

Green vegetables	670 g
Turnips and swedes	400 g
Skimmed milk	0·3 l
Full-rich milk	0·2 l
Potatoes	150 g
Hens' eggs	60 g
Jam	50 g
Meat	50 g
Cheese	45 g
Bread	42·5 g
Legumes and pastas	30−40 g
Sugar	25 g
Butter or margarine	13·5 g

To illustrate the energy content of some of the most important foodstuffs, Table 26 gives the weight or quantity of each which is equivalent to 420 kJ.

Needs of 'sedentary workers'

For 'sedentary workers' as well as for the great majority of female occupations, it is broadly true that *the quantity of food should be restrained in favour of high quality: in other words, fewer kJ, and more vitamins, minerals and trace elements.*

These categories of workers would also be well advised to cut down on energy-rich and highly refined foodstuffs, and give preference to natural foods containing protective elements: vegetables, salads, raw fruit, milk, brown bread and liver.

Under normal circumstances a person takes in just enough food to supply the energy he or she needs, regulating this according to feelings of hunger, and so achieving an energy balance. Disturbances of this balance are fairly common among sedentary workers, who have a tendency to eat more than they need for their everyday life. Such people are often visibly overweight.

Needs of manual workers

Manual workers have quite different problems. They need a diet that is energy-rich, but not bulky, and so tend to prefer food that is protein-rich and fatty. Carbohydrate foods are bulky and full of indigestible roughage. They include all kinds of sugar (sweet and non-sweet), which make up the greater part of flour, potatoes, pastas, and of course all sweet foods.

If a heavy worker wants to take in 15 000 kJ in the form of potatoes, he will have to eat 5 kg of them. It would certainly be wrong for a manual worker to try to make up the balance of energy that he needs from carbohydrate foods alone. Such a diet would be too bulky and would overload the digestive organs. The recommended course is for manual workers to increase their intake of proteins and fats to approximately double the normal value. For a man weighing 70 kg this means about 100—110 g of protein per day, and about the same amount of fat.

Muscular work requires increased amounts of vitamin B_1 and phosphates. A heavy worker should have the benefits of brown bread (i.e., rye or wholemeal) and milk products. On the whole the energy-rich diet of manual workers will provide enough of the protective vitamins and minerals. To summarize, manual workers should have an energy-rich diet, giving preference to meat, eggs, milk, butter, cheese and brown bread.

The importance of proteins and fats

Since protein of animal origin is more valuable than plant protein for body building and muscular strength, half of the intake should come from meat, eggs and milk. A 70 kg man could find the necessary 50 g of animal protein in 1·5 l of milk, 300 g of meat, or seven eggs.

Fats are the foodstuffs that are richest in energy: 100 g of fat can replace about 300 g of bread or even 1 kg of potatoes. A further advantage of fat for heavy workers is that it remains in the digestive organs longer, and so postpones the onset of hunger.

Generally speaking, too much fat is eaten in developed countries. Many studies have shown that ailments of the heart and circulation can be partly traced back to too much fat in the diet. The risk is less to manual workers than to 'pen-pushers', who are strongly recommended to avoid fats and greasy foods as much as possible. The origin of the fats eaten is also of some significance to health, and one half of the fats should be plant oils, and the other half animal fats. Milk and milk products should be the preferred forms of animal fats, since they contain many vitamins and minerals.

Overall working time and nutrition

'Continental' working hours, with a long two-hour break at midday, allowed the worker to go home to take the midday dinner with the family, and still have time for a rest, if he or she wished. Today nearly all industries and most services have adopted the short lunch break and an earlier release in the evening.

A shorter midday break usually means eating in a works canteen or a nearby restaurant. A meal away from home is more expensive and there is less time for eating, and a worker instinctively feels that it would be unwise to go back to work immediately after a large meal; hence a change in eating habits, transferring the main meal to the evening, and making the midday meal little more than a large snack. This has the advantage that the afternoon's work is less affected by digestive problems.

From a medical point of view it should also be noted that *a midday break of 45—60 minutes is usually enough for relaxa-*

tion, provided that there is also another rest pause of 10−15 minutes, both morning and afternoon, for relaxation and eating a snack.

Food intake and biological rhythm

In a Swedish gas works the reading errors of several inspectors were tabulated daily over a period of 19 years. Figure 132 shows these reading errors, according to the time of day at which they occurred. It is clear, among other things, that they reached a maximum immediately after the midday break, that is while digestion was still going on

This observation confirms a biological law, by which loading the digestive system damps down the state of readiness of the entire organism. This effect of large meals has been known for a long time, and finds expression in the familiar proverb: 'A full stomach makes a lazy student'.

There are several old studies showing the effects of the distribution of meal times. A classic study was carried out by Haggard and Greenberg [100] who observed that small snacks every two hours kept the blood sugar and efficiency at a higher level throughout the working day. These results were confirmed in a tennis-shoe factory where three meals plus two snacks were associated with a higher productivity than conditions with fewer meals and snacks. These results, as well as more recent studies, lead to the conclusion *that the intake of food five times daily (three meals and two snacks) is good for both health and efficiency, and is to be recommended, provided it doesn't lead to excessive energy intake.* This recommendation applies particularly when continuous working with

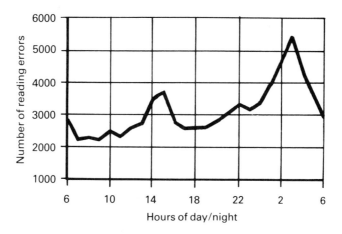

Figure 132. Reading errors of gas meter inspectors, according to the time of day.
Evaluation of 175 000 records, with a total of 75 000 errors, between 1912 and 1931. Besides the distinct peak after the midday meal, mentioned in the text, there is a much higher peak in the early hours of the morning, when readings are taken in the dark. After Bjerner et al. [18].

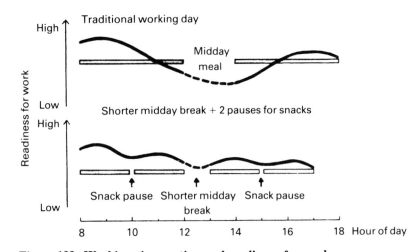

Figure 133. Working time, eating and readiness for work.
The dotted blocks cover the working periods, and the curves indicate the conjectural rise and fall of readiness for work. During the working day there should be two breaks for snacks, in addition to a short midday mealtime (below) when the curve does not fall as low as it does with a larger midday break (above).

a short midday break is in operation, since then the need for extra periods to relax and eat is even greater. Figure 133, based on current scientific knowledge, and general experience, shows a 'theoretical' curve of eagerness for work, both during traditional working hours and during a working day that is planned according to physiologically sound principles.

Intake of fluids   A person needs not only food, to provide energy, but also water, to maintain a correct water balance. The average requirement is 35 g water for every kilogram of body weight per 24 hours (= 2−2·5 l per day). The water that is drunk and that which is obtained from the food eaten is continuously excreted through the kidneys and the sweat glands. The liquid excreted is not pure water, but a fluid rich in waste materials (urea, sodium chloride and many other metabolic end-products). Although foodstuffs have a high fluid content (e.g., meat 70−80%; bread 43%; fruit 85%; potatoes 78%; pastas 19%), we still need more liquid. The amount varies between individuals, but is generally about 0·5−1 litre, rising to 1·5−2 litres and more on a hot summer day.

The amount of water we drink is governed by feelings of thirst, which in turn depend mainly on the concentration of salts in the blood. An increase in salt concentration makes one more thirsty.

In summer much of the water is lost by sweating. Since sweating is essential to maintaining body temperature, like the cooling of an engine, the consequent water loss must be replaced in summer, in tropical countries, and in hot jobs in industry. This is best done by drinking easily assimilated liquids such as tea, coffee or soft drinks. The loss of minerals during heavy sweating may be offset by taking salt tablets.

Summary of snacks

Snacks between meals are important both for sedentary and manual workers, to meet a significant part of their daily fluid requirements, as well as of additional energy-producing food, according to their bodily needs. Table 25 shows the necessary energy content of such between-meal snacks, but people vary greatly in their eating habits, and hence in their choice of snacks.

The range of choice lies between a refreshing drink that is low in kJ and, at the opposite extreme, a heavy snack equivalent to some 1600 kJ, with many intermediates. A few suggested snacks between meals are listed in Table 27, the choice depending primarily on the fluid and energy requirements of the person concerned.

A drink with few kJ (soup, tea, or coffee) and no solid food is recommended for desk workers, whereas manual workers need an energy-rich drink such as ovaltine, apple juice, milk or yoghurt, supplemented by bread and cheese, sausage or fruit.

Coffee and tea are especially popular drinks as snacks because they have an immediate stimulating effect, though this is slight and does not last very long. The worker feels a need for frequent stimulants of this kind, and there is no objection on medical grounds, provided that they are not drunk to excess. A certain amount of stimulation is a good thing where the work is monotonous, yet responsible, as often happens nowadays with such jobs as a switchboard operator, or a controller of a machine that is mainly automatic.

Table 27. Suggestions for between-meal snacks, and their energy content.

Type of snack	Number of kJ
1 cup of mineral water	—
1 cup of soup	40−60
1 cup of tea with two lumps of sugar	150
1 cup of coffee with milk and two lumps of sugar	155
1 cup of apple juice	270
1 cup of milk or yoghurt	275
1 cup of ovaltine in milk	540
Bread (50 g)	500
Bread with fruit	1000
Bread with cheese	1250
Bread and sausage	1250

Effects of snacks on teeth

An important aspect of between-meal snacks is the concern with dental health, because the link between daily sugar intake and the incidence of dental caries is undeniable. The stickiness and physical consistency of sugar foods is bad for the teeth, especially pastries, nutty crunch, and many kinds of sweet biscuit, chocolate, bananas and crystallized fruit. On the other hand bread and fruit are rarely harmful, in spite of their starchiness. From a dental point of view the following are recommended items for between-meal snacks: *apples, nuts, fresh fruit, mineral water or skimmed milk, bread and butter, cheese, yoghurt, sausage, meat.*

16. Nightwork and shiftwork

16.1. Day- and night-time sleep

The problem

In recent times all industrial countries have turned over more and more to continuous production. This is why shiftworking is no longer a fringe problem, but of ever-increasing importance.

The main reasons for going over to continuous production are economic ones. Many manufacturing processes are said to be economic only with this form of working. In other cases expensive machinery needs to be used 24 hours a day to be profitable.

We have already said that the human organism is in its ergotropic phase (geared to performance) in the daytime, and in its trophotropic phase (occupied with recuperation and replacement of energy) during the night. Hence the nightworker approaches his work, not in a mood for performance, but in the relaxed phase of the daily cycle. Herein lies the essential physiological and medical problem of nightwork. Another aspect is the burden it puts on family life, and the social isolation. Ergonomics is therefore faced with the problem of planning work schedules in such a way that shiftwork does as little harm as possible to health and to social life.

Comprehensive surveys of the problem of nightwork and shiftwork may be found in the publications of Mott [189], Swensson [250], Rutenfranz and Knauth [225] Andlauer *et al.* [7], Colquhoun [48], and Conroy and Mills [49].

The circadian rhythm

The various bodily functions of both man and animals fluctuate in a 24-hour cycle, called the circadian rhythm (*circa dies* = approximately a day).

Even if the normal influences of day and night are excluded, e.g., in the Arctic, or in a closed room with unchanging artificial lighting, a kind of internal clock comes into play, the so-called endogenous rhythm. This varies in different individuals, but usually operates a cycle of between 22 and 25 hours.

Under normal conditions endogenous circadian rhythms are synchronized into a 24-hour cycle by various '*time-keepers*':

217

Table 28. Circadian bodily functions that increase by day and decrease by night.

Body temperature
Heart rate
Blood pressure
Respiratory volume
Adrenalin production
Excretion of 17-keto-steroids
Mental abilities
Flicker-fusion frequency of eyes
Physical capacity

---

1. Changes from light to dark and vice versa.
2. Social contacts.
3. Work.
4. Knowledge of clock time.

The bodily functions that are most markedly circadian are sleep, readiness for work, and many of the autonomic, vegetative processes such as metabolism, bodily temperature, heart rate and blood pressure. Table 28 summarizes a few characteristic day/night changes.

Effect of circadian rhythms

The bodily functions listed above certainly show these trends throughout the 24 hours, but they do not all reach their maxima and minima at the same time. There is a distinct phase-difference among them. Taken as a whole they confirm the rules mentioned above:

1. During daytime all organs and functions are ready for action (ergotropic phase).
2. At night most of these are damped down, and the organism is occupied with recuperation and renewal of its energy reserves (trophotropic phase).

Normal sleep

The most important function that is geared to circadian rhythm is sleep. While it is still not possible to say just what is the scientific function of sleep, it can certainly be said that sleep that is undisturbed either in quantity or in quality is a prerequisite for health, well-being and efficiency.

An adult human being requires about eight hours' sleep per night, although there are considerable individual variations. While there are some people who need ten hours' sleep, if they are to be fresh and alert, others need only six hours, or even less. It was said of Thomas Alva Edison that he needed only about three hours, and that he himself dismissed even these three hours as merely a bad habit.

Length of sleep is mainly a matter of age. A new-born child needs 15−17 hours daily during its first six months, whereas old people sleep less and less, and often this is in broken periods.

The quality of sleep is not uniform, but cyclical, and has various stages of different depth (see Figure 134). The following five stages may be distinguished.

Stage 1. The electroencephalogram (EEG) shows low amplitudes, with many theta waves. This is the stage of going to sleep, sleeping lightly. Duration 1−7 minutes.

Stage 2. EEG shows low amplitudes. Besides the theta waves there are also the so-called 'sleep spindles', strong peaks between 12 and 14 Hz, following in quick succession. Stage 2 is a condition of light sleep and its duration is about 50% of total sleeping time.

Stage 3. EEG shows increased amplitudes and a decrease in frequencies, up to 50% of the waves being below 2 Hz. Many delta rhythms, interspersed with sleep spindles. Deeper sleep.

Stage 4. More than 50% of the waves in the EEG are below 2 Hz. Maximum synchronization and deepest phase of sleep.

Stage 5. Rapid eye movements (REM). In EEG similar to Stage 1, with a mixture of frequencies. REM − frequent salvoes of quick movements of the eyes − are characteristic, this is the stage during which dreams are especially common. Despite the picture on the EEG, the REM stage is characterized by maximum relaxation of the muscles and resistance to being awakened; hence Stage 5 is known as the 'paradoxical stage'.

Quality of sleep   Although little is yet known about the significance of the five stages of sleep, it can be said in general terms that *Stages 3, 4 and 5 are the ones that have particular recuperative properties.* These stages determine the quality of one's sleep.

As mentioned above, cyclical changes take place during sleep, with about four descents into deep sleep, linked by intervening shallow periods. Figure 134 shows the cyclical course of an ordinary night's sleep.

Daytime sleep of night workers   For a long time, company medical advisers have recorded frequent cases of disturbed daytime sleep among night workers. Part of this disturbed sleep must be attributed to noise, which is usually greater in a residential area during the day

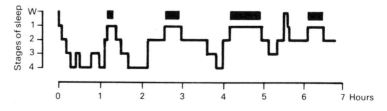

Figure 134. The cyclic course of a night's sleep.
W = being awake. Black bars = periods of rapid eye-movements (REM).

than at night, but many nightworkers say in addition that they feel a certain restlessness during the day, and their daytime sleep is not refreshing enough [225].

Length and
quality of
daytime sleep

Some recent EEG studies of length and quality of sleep among night-shiftworkers are of interest. Figure 135 shows the results of a study by Françoise Lille [167] who analysed in detail the daytime sleep of 15 regular night-shiftworkers.

It appeared that daytime sleep was distinctly shorter than the night sleep the workers took on their rest day. The average length of sleep in the daytime was 6 hours, whereas on the rest day the average varied between 8 and 12 hours, with longer sleep on the second of the two rest days than on the first. Françoise Lille concluded that *the nightworker accumulated a 'sleep debt' which was 'paid back' on the two rest days.* Evidently a single day's rest was not enough for this purpose.

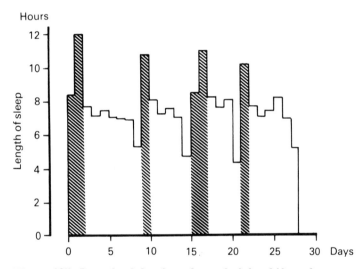

Figure 135. Length of daytime sleep of night-shiftworkers.
White columns = total of daytime sleep. Shaded columns = night sleep on rest days. Average of 15 workers. After Lille [167].

Daytime sleep on
the EEG

Detailed analysis of EEGs showed the quality, as well as the duration, of daytime sleep was impaired, as evidenced by a greater number of periods of light sleep, and more body movements. These findings were later confirmed by other authors.

Comparison between sleepers in noisy surroundings and in a soundproof room showed that the disturbance was not caused by noise, but was an integral feature of daytime sleep.

To summarize, all these studies show that sleep following a night-shift is curtailed and of little restorative value.

Capacity for work at night

Both mental and physical working capacity show a characteristic circadian rhythm. As an ergonomic example the reading errors of the Swedish gas inspectors may be mentioned once again [18]. As is evident from Figure 132 psychophysiological readiness for work is at a maximum in the morning, and in the second half of the afternoon, whereas it is poor immediately after the midday break and declines even more at night.

One more example may be given, coming from some work by Prokop and Prokop [212], during which 500 truck drivers were asked at what times of day they had fallen asleep at the wheel at least once. The numbers of replies for each of the hours is shown in Figure 136.

The statements about falling asleep show a clear daily cycle, with one peak in early afternoon and an even more pronounced peak during the night. These examples, to which others could be added, show that readiness for action is high during the daytime and declines at night. These results reflect the rule that was formulated in Chapter 15 that the human organism is performance-orientated during the daytime and damped down at night.

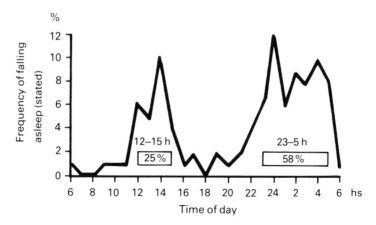

Figure 136. Frequency with which 500 truck drivers fell asleep at the wheel, in relation to the time of day.
After Prokop and Prokop [212].

Productivity and frequency of accidents

These facts led to the assumption that nightwork would be conducive, not only to lower output, but also to more frequent accidents. Several authors have recorded these, yet the hypothesis is still not confirmed. Often the accident rate at night seems scarcely altered, or even reduced [6]. This contradiction between theory and practice is hardly to be wondered at, if we think of the conditions surrounding the nightworker (fewer disturbances from other people, higher wages, etc) compared with those of the dayworker. Moreover, there is much evidence that the nightworker has often made a 'positive choice' of particularly rewarding work.

Reversal of circadian rhythm

It has already been pointed out that the circadian rhythm is affected by a variety of other time-keeping factors. It is reasonable to assume that the circadian rhythm of the night-worker may be reversed by the 'work' factor. So far this assumption has only been partially confirmed.

Several studies [48, 49, 76, 144] have been carried out, mostly recording body temperature and heart rate, to try and analyse the reversal of circadian rhythms. All authors stress that this reversal is not complete even after several weeks, that the curves become flattened, yet the maxima can hardly be said to be interchanged in position. So we conclude that *biological circadian rhythms show the first signs of reversal after several night-shifts, but the reversal is not usually total, even after several weeks.*

16.2. Nightwork and health

It has been known for a long time that nightworkers commonly have bad health, and that more and more workers are forced to give up shiftwork for this reason. When shiftwork increased after World War II these matters of adjustment became a problem in industrial medicine, which was analysed in detail in several large field studies.

Sickness rate

The first big surveys came from Scandinavia. In Norway Thiis-Evensen [255] and Aanonsen [1] studied overall sickness rates among 6000 and 1100 workers, respectively; see Table 29.

Table 29. Sickness rates among shiftworkers in Norwegian factories. *Studies made between 1948 and 1959 by Thiis-Evensen [255] and Aanonsen [1]. The percentages quoted relate to the total number of workers in each group studied.*

	Thiis-Evensen		Aanonsen		
Ailments	Day work	Night work	Day work	Night work	Former night workers
Stomach troubles	10·8	35·0	7·5	6·0	19·0
Ulcers	7·7	13·4	6·6	10·0	32·5
Intestinal disorders	9·0	30·0	11·6	10·2	10·6
Nervous disorders	25·0	64·0	13·0	10·0	32·5
Heart troubles	—	—	2·6	1·1	0·8

'Positive choice' of nightworkers

Thiis-Evensen's survey showed that shiftworkers had significantly more digestive ailments and nervous disorders, and

Aanonsen's work revealed an interesting corollary. Among the dayworkers investigated there were many who had abandoned shiftwork, either on health grounds, or because they did not like it. This group of former nightworkers, who had certainly exercised a 'negative choice', showed a distinct increase in digestive and nervous troubles. This discovery proved that comparisons with so-called normal groups must be carried out with caution, and that up to a point, night-shiftworkers must be regarded as being a 'positive selection' of particularly tough workers. This might also account for the contradictory results of surveys of sickness rates.

In spite of these statistical difficulties, an ever increasing sickness rate has been observed during the past 20 years among 'active' as well as former night-shiftworkers. Several reports should be mentioned here [8], according to which night-shiftworkers often misuse drugs, taking stimulants during the night and sleeping tablets during the day. The reasons for the increased liability to nervous disorders and ailments of the stomach and intestines are primarily:

1. *Chronic fatigue.*
2. *Unhealthy eating habits.*

Occupational sickness among nightworkers

Nowadays, it is justifiable to talk about *occupational sickness* among nightworkers, the dominant symptoms being those of *chronic fatigue* (see also Chapter 11), thus:

1. *Weariness, even after a period of sleep.*
2. *Mental irritability.*
3. *Moods of depression.*
4. *General loss of vitality, and disinclination to work.*

The state of chronic fatigue is accompanied by an increased liability to psychosomatic disorders, which in nightworkers commonly takes the form of:

1. *Loss of appetite.*
2. *Disturbance of sleep.*
3. *Digestive troubles.*
4. *Stomach and duodenal ulcers.*

Thus the nervous disorders observed among nightworkers are no more than symptoms of chronic fatigue, which, combined with unhealthy eating habits, is the cause of the increased liability to digestive troubles.

The causes

What, then, are the actual causes of occupational sickness among nightworkers? The answer to this question lies in what we have already said about the circadian rhythm and the disturbances arising from the change from daywork to the night-shift. A conflict is generated in the body of a nightworker by 'desynchronization' of his time-keeping mechanism;

Fitting the task to the Man

the 'working' cycle is opposed to the 'light—dark' and 'social contact' cycles. None of these mechanisms seems to be fully dominant, so that the functional unity of the body is lost, and the harmonic correlation between the separate biorhythms is impaired.

Symptoms

Expressed in other words: *since a complete adjustment to nightwork does not take place quickly enough, the night-worker's bodily system is only partly switched over to 'working' at night, and 'sleeping and resting' by day. The result is insufficient sleep, both in quantity and quality, with less adequate recuperation, resulting in chronic fatigue with its associated symptoms.*

Since at the same time eating habits are unhealthy (meals at unfamiliar times and so on), the psychosomatic symptoms tend to show themselves mainly in digestive disorders.

The nature of occupation sickness of nightworkers, with its causes and symptoms, is set out diagrammatically in Figure 137.

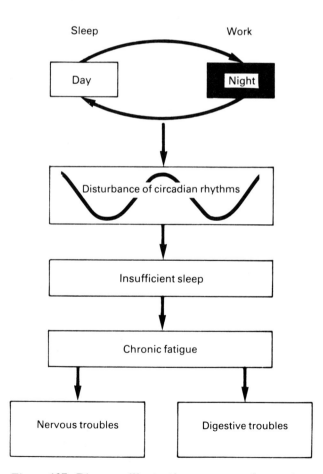

Figure 137. Diagram illustrating causes and symptoms of occupational ailments among shiftworkers who periodically work at night.

Individual susceptibility

These ailments do not afflict every worker in the same way, and even if they show the same symptoms, the extent of the disorders varies very much from one person to another. *It is broadly true that about two-thirds of shiftworkers suffer some degree of ill-health, and about one-quarter sooner or later abandon shiftwork because of major health problems.*

Effect of age

Resistance to the special stresses of nightwork declines with age. Most burdensome is the need to change over to being active during the sluggish period of the night. The older worker is less adaptable, and tires more easily. On the other hand the older worker cannot enjoy the benefit of longer sleep since the sleep of older people is very easily disturbed. *Hence the older worker suffers both from greater stresses and fewer opportunities to recuperate from them.* In fact many surveys have shown that shiftworkers in age groups over 40 are distinctly more prone to disturbed sleep and complain of their ill-health [9]. It may be concluded that *during the course of a year people do not usually become accustomed to nightwork, but on the contrary they become increasingly liable to occupational ailments.*

Social aspects of shiftwork

Since 'social well-being' is closely related to physical health, we must now briefly consider the social effects of shiftwork.

In the forefront of these is the disruption to family life, interference with wider social contacts among friends, and fewer opportunities to participate in group activities.

Many investigations show that most of the people questioned give priority to complaints about fewer meal times with their families. As a result, the father of the family is forced to leave the upbringing of the children to his wife.

Free time

Another complaint that is frequently expressed concerns disruption to the social life outside the family circle. Active participation in group activities, whether in sport or politics, is so limited that the nightworker often feels that he is excluded from society altogether. There are similar impediments to the cultivation of friendships, especially if there are not many other shiftworkers living nearby. This state of affairs also dictates what the shiftworker can do in his free time, so that he often pursues solitary hobbies. Some authors talk in this context of a tendency for shiftworkers to 'feel themselves to be on the fringe of society' or of 'social isolation'.

Opinion polls of shiftworkers

Opinion polls have shown that many people are in two minds about shiftwork. On the one hand they are opposed to it on grounds of health and social life, but on the other they see certain advantages in it, such as more pay, or more freedom to plan their leisure. On the whole, however, the drawbacks predominate.

The three shifts

Each of the three shifts has its advantages and drawbacks.

The early shift makes a communal evening meal possible. Leisure activities are possible, either in the afternoon or evening. If the shift begins very early (e.g., 4 am), it is tiring and night sleep is cut short.

The late day-shift is particularly bad for social life since there are no opportunities for family life, either round the table, or in leisure activities. On the other hand sleep is good after this shift.

The night-shift is bad from all angles. Family life is often limited to taking the evening meal together. Leisure activities are usually possible only in the second half of the afternoon. Sleeping habits vary: one group of nightworkers interrupt their daytime sleep for a midday meal, and then lie down again afterwards; the others sleep through until early afternoon. The night-shift is bad for social life, and very tiring because all sleeping is carried out in the daytime.

16.3. Organization of shiftwork

Distribution of shifts during the day

On the three-shift system the day is divided into three equal periods of eight hours each. A common system is this:

Early shift:	0600−1400 hours
Late day-shift:	1400−2200 hours
Night-shift:	2200−0600 hours

But there are many variants. In the USA for example, 8−16−24 hours is commonly worked, and this arrangement seems to have advantages, both physiological and social. Each shift allows at least one mealtime in the family circle, and at the same time provides good opportunities for sleeping, especially on the early and late day-shifts.

A few firms work a system of two shifts of 12 hours each, but a 12-hour working day cannot be recommended from the standpoint of either industrial medicine or ergonomics (see Chapter 15). At most, exceptions might be made for undemanding jobs, with long built-in pauses. When shifts are as long as this, each shift, whether day or night, is followed by two rest days, and many workers like this.

Rotation of shifts

In Europe periodic rotation of shifts is the rule, but in the USA it is not uncommon to work the same shift all the year round. Mott [189] sees certain social advantages in this arrangement, but in the long run continuous nightwork is not acceptable, either on social or medical grounds, at least in big doses.

Shift rotation cycle

Until 1960, many experts had been of the opinion that the intervals between shift rotation should be as long as possible. Recommendations for rotation every 3–4 weeks were based on the idea that people need several days to change their biological rhythm, and adaptation to the new shift can take place only if several weeks are allowed. Nowadays we know that this interpretation is misleading. Even after several weeks, adaptation is not complete, especially with regard to sleep, one of the most important bodily functions. The day-time sleep of workers on the night-shift remains inadequate, both quantitatively and qualitatively for a long time.

The latest information leads to the recommendation that rotation of shifts should be short-term.

Criteria for shift rotation

As a start, it may be helpful to consider what criteria apply to shift systems.

The following may be considered the most important requirements:

1. Loss of sleep should be as little as possible, so as to minimize fatigue.
2. There should be as much time as possible for family life and other social contacts.

The best shift plans to meet these requirements are those with single, isolated night-shifts, each followed immediately by a full 24-hours rest. Figure 138 shows a shift plan which meets most of the requirements.

From this it can be seen that over a period of four weeks there is only one set of three consecutive night-shifts. All the other night-shifts are scattered singly, and each is followed immediately by a rest day. A very good feature of this plan is the distribution of free shifts, which, throughout the year, include 13 complete weekends, Saturday to Monday inclusive.

Two plans widely used in England are the 2–2–2 system (the so-called 'metropolitan rota') and a 2–2–3 system that is called the 'continental rota'. Both of them are short rotations, which comply with the current ergonomic recommendations. The two are shown in Figures 139 and 140.

It will be seen that on one system the free days follow two nights' work, and on the other three nights. The 2–2–2 system is slightly the less favourable, because a free weekend (Saturday/Sunday) comes only once in eight weeks. The 2–2–3 system is more advantageous in this respect because a free weekend occurs every four weeks.

Short-term rotations are made more difficult because they sometimes bring production to a halt at weekends, but it should be possible to reach a compromise over this, as suggested by Rutenfranz and Knauth [225].

Mon	Tue	Wed	Thur	Fri	Sat	Sun
N	—	E	L	N	—	—
—	E	L	N	—	E	E
E	L	N	—	E	L	L
L	N	—	E	L	N	N

Weekend patterns and frequency per year			
Saturday	Sunday	Monday	Frequency per year
—	—	—	13
E	E	E	13
L	L	L	13
N	N	N	13

Figure 138. Top: an example of a shift rota, in which the night shifts are widely scattered. Bottom: summary of free shifts (rest periods) over the year.
E = early shift; L = late day shift; N = night shift.

	M	T	W	Th	F	S	Su		M	T	W	Th	F	S	Su
1. Week	E	E	L	L	N	N	—	5. Week	N	N	—	—	E	E	L
2. Week	—	E	E	L	L	N	N	6. Week	L	N	N	—	—	E	E
3. Week	—	—	E	E	L	L	N	7. Week	L	L	N	N	—	—	E
4. Week	N	—	—	E	E	L	L	8. Week	E	L	L	N	N	—	—

Figure 139. The 2-2-2 shift system ('metropolitan rota').
E = early shift; L = late day shift; N = night shift.

	M	T	W	Th	F	S	Su		M	T	W	Th	F	S	Su
1. Week	E	E	L	L	N	N	N	3. Week	N	N	—	—	E	E	E
2. Week	—	—	E	E	L	L	L	4. Week	L	L	N	N	—	—	—

Figure 140. The 2-2-3 shift system ('continental rota').
E = early shift; L = late day shift; N = night shift.

16.4. Recommendations

Shiftwork that includes night-shifts is burdensome, and often leads to ill-health which can rightly be classified as occupational. *Nightwork is therefore a danger to health.*

Since there is no way of planning shiftwork that significantly reduces this occupational risk, it should be introduced only with the greatest hesitation.

Rutenfranz [225] has written in this context: 'In my opinion, and from the standpoint of medical safeguards in industry, continuous production is permissible only where it is unquestionably essential to the manufacturing process. Its introduction simply to increase profits is to be deplored.' *If a night-shift is unavoidable, then the following nine recommendations should be considered:*

1. *Night-shiftworkers should not be engaged when they are below 25 years old, or over 50.*
2. *Workers with a tendency to ailments of the stomach and intestine, who are emotionally unstable, prone to psychosomatic symptoms and to sleeplessness, should not be employed on nightwork.*
3. *Workers who live by themselves, who live far away from their work or in noisy neighbourhoods are unsuitable for nightwork.*
4. *The usual three-shift system, changing over at 6−14−22 hours would be better altered to 7−15−23 or 8−16−24 hours.*
5. *Short-term rotations are better than long-term ones, and continuous nightwork without change should be avoided.*
6. *A good shift rotation either calls for scattered single nights at work, or else the 2−2−2 or 2−2−3 rotation.*
7. *Whether one, two or three nights are worked in a row, they should be followed immediately by at least 24 hours' rest.*
8. *Any shift plan should include some weekends with at least two consecutive rest days.*
9. *Every shift should include one break for a hot meal, to ensure adequate nourishment.*

17. Vision

17.1. The visual system

Visual perception   The eyes, acting as receptor organs, pick up energy from the outside world in the form of light waves and convert these into a form of energy that is meaningful to a living organism, i.e., into bioelectric nerve impulses. It is only through the integration of the retinal impulses by the brain that we have visual perception. If the afferent sensory nerves, linking the eyes with the brain, are cut, we become blind. Perception in itself does not give a precise image of the world outside: our impressions are a subjective modification of what is perceived. Thus:

> A particular colour seems darker when it is seen against a bright background as opposed to a darker one.
> A straight line appears distorted against a background of curved or radiating lines.
> Individual variations in the interpretation of sense data may be critical in certain situations. People differ in experience, attitude and preconceived ideas.
> People differ greatly in the intensity with which they react to sensory data.

Control mechanisms   The successive stages of seeing can be simplified as follows: light rays from an object pass through the pupil aperture and converge on the retina. Here the light energy is converted into the bioelectric energy of a nerve stimulus which then passes as a nerve impulse along the fibres of the optic nerve to the brain. At a first series of intermediate nerve cells — called neurones — new impulses are generated which branch off to the centres which control the eyes, regulating the width of the pupil, the curvature of the lens and the movements of the eyeball. These control mechanisms keep the eyes continuously directed at the object, and this is automatic, not under conscious control. At the same time the original sensory impulses travel further into the brain, and after various filtering processes end up in the cerebral cortex, the seat of consciousness. Here all the signals coming from the eye are integrated into a picture of the external world. Here, too, new impulses arise which are

The visual apparatus

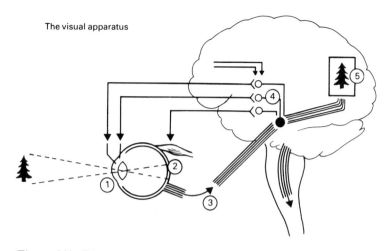

Figure 141. Diagram of the visual system.
1 = cornea and lens. 2 = light received on the retina. 3 = transmission of optical information along the optic nerve to the brain. 4 = synapses and feedback to the eye. 5 = visual perception of the external world in the conscious sphere of the brain.

responsible for coherent thought, decisions, feelings and reactions. These processes of the visual system are shown in Figure 141.

In reality, the essential processes of vision are nervous functions of the brain; the eye is merely a receptor organ for light rays. The complete visual system controls about 90% of all our activities in everyday life. It is even more important in a great many jobs. If the numerous nervous functions that are under stress during seeing are considered, it is not surprising that the eyes are sometimes an important source of fatigue.

The eye

The eye has many elements in common with a photographic camera: the retina corresponds to the light-sensitive film, and the transparent cornea, the lens and the pupil, with its

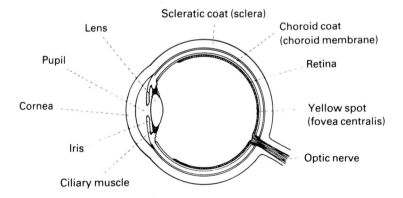

Figure 142. Diagrammatic horizontal section through the right eye.

variable aperture, represent the optics of the camera. Cornea and lens together refract the incoming rays of light and bring them to a focus on the retina mainly in the fovea centralis.

The principal parts of the human eye are shown in Figure 142.

The retina

The actual receptor organs are the visual cells embedded in the retina, consisting of 'cones' for daylight vision and of highly sensitive 'rods' for vision in dim light. The visual cells convert light energy by photochemical reactions into nervous impulses which are then transmitted along the fibres of the optic nerve.

The fovea

The human eye contains about 130 million rods and 7 million cones, each of which is approximately 0·01 mm long and 0·001 mm thick. On the posterior surface of the eye, a few degrees on either side of the optical axis, is the retinal pit, or *fovea centralis*, characterized by a thinner covering than the surrounding area. The thin covering allows the light rays to pass directly to the visual cells, which, in the fovea, consist entirely of cones, here at their maximum density of about 10 000 cones per mm². Each foveal cone has its own fibre connecting it to the optic nerve. For these reasons the fovea has the highest resolving power of any part of the retina, up to about 12 seconds of arc. Since vision is most acute in the area of the fovea, it is instinctive to look at an object closely by turning the eye until the image falls onto this area of the retina which is called the area of central vision. Any object or sign that is to be seen clearly must be brought to this part of the retina, which covers a visual angle of only 1°.

Rods and cones

Outside the foveal area the cones are considerably fewer, and one nerve fibre serves several rods and cones. Here the rods are distinctly more abundant than cones, and they become more numerous as the angle from the fovea increases, whereas the number of cones declines. Although rods are more sensitive to light than cones they do not detect such fine differences of either shape or colour. The rods are the more important light-detecting organs in poor visibility and at night.

The sharp picture

So only objects focused on the fovea are seen clearly, while other images become progressively less distinct and blurred as distance from the fovea increases. Normally the eye moves about rapidly, so that each part of the visual field falls on the fovea in turn, allowing the brain to build up a sharp picture of the whole surrounding.

The visual field

The visual field is that part of one's surrounding that is taken in by the eyes when both eyes and head are held still. Only objects or signs within a small cone of 1° apex are focused sharply. Outside this zone objects become progressively more

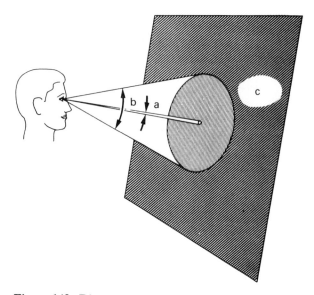

Figure 143. Diagram of the visual field.
a = zone of sharp vision; angle of view of 1°. b = middle field: vision unsharp; angle of view from 1° to 40°. c = outer field: movements perceptible; angle of view from 41° to approximately 70°.

blurred and indistinct. If the eyes are kept still when reading only a few letters can be focused. More details about the physiology of reading will be discussed later in this Chapter.

As shown in Figure 143, the visual field can be divided up as follows:

Area of distinct vision:	viewing angle 1°
Middle field:	viewing angle 40°
Outer field:	viewing angle 40−70°

Objects in the middle field are not seen clearly, but strong contrasts and movements are noticed: alertness is maintained by quickly shifting the gaze from one object to another. The outer field is bounded by the forehead, nose and cheeks; objects in this area are hardly noticed unless they move.

17.2. Accommodation

Accommodation means the ability of the eye to bring into 'sharp focus' objects at varying distances from infinity down to the nearest point of distinct vision, called the 'near point'. If we hold up a finger in front of the eye, the finger can be focused sharply, leaving the background blurred, or the background can be focused sharply, leaving the finger indistinct. This demonstrates the phenomenon of accommodation.

An object is seen clearly only when refraction through the cornea and lens produces a tiny but sharp image on the retina, the three components forming an optical system. Focusing on near objects is achieved by changing the curvature of the lens, by contraction of *the muscles of accommodation*, called *the circular ciliary muscles*.

Distant objects   When the ciliary muscles are relaxed, the refraction of cornea and lens is such that parallel rays from distant objects are focused onto the retina. Therefore, when attention is allowed to wander over distant objects, the eyes are focused on 'infinity' and the ciliary muscles remain relaxed.

Resting accommodation   For a long time it was assumed that accommodation focused on infinity was also the resting position of the eye. But several studies revealed that in the dark the resting position corresponds to focusing distances lying somewhere between the near point and infinity. Krueger and Hessen [153] determined a mean focusing distance of 800 mm for students in a resting position. This distance seems to move gradually towards 'infinity' as we get older.

Near vision   Without accommodation the image of an object nearer to the eye would fall behind the retina, which would receive a blurred impression. To avoid this the ciliary muscle increases the curvature of the lens so that the image is brought back into the plane of the retina. In near vision the lens is continuously adapting the focal length in such a way that a sharp image is projected onto the retina. To maintain focus on a near object the ciliary muscle must continuously exert a contracting force.

The accommodated lens is in constant motion. When viewing a target the lens will oscillate in a certain range at a rate of about 4 times per second. Even when reading a book the lens remains quite active. It seems that these movements of the lens and the perception of blur are important for the automatic regulation of accommodation. The key to comfortable viewing is accommodation; it means that the image is well focused on the retina.

After viewing a near object for some time the lens may not immediately return to its relaxed position. This condition, referred to as 'temporary myopia', may remain for several minutes.

The near point   As already mentioned the shortest distance at which an object can be brought into sharp focus is called the *near point* and the furthest away is the *far point*. The nearer the object is focused the greater is the load on the ciliary muscle. The near point is a measure of the power of the ciliary muscle and of the elasticity of the lens. It moves further away as the ciliary muscle becomes tired after a long spell of close work. Many experiments have shown that prolonged reading under unsuitable conditions is associated with an increase in the near point distance, a phenomenon considered a symptom of visual fatigue.

Age and accommodation

Age has a profound effect on our powers of accommodation, because the lens gradually loses its elasticity. As a result the near point gradually recedes, whereas the far point usually remains unchanged or becomes slightly shorter.

The average distance of the near point at various ages is reported in Table 30.

Table 30. Average near point distance at different ages.

Age (years)	Near point (mm)
16	80
32	120
44	250
50	500
60	1000

Presbyopia

When the near point has receded beyond 250 mm close vision becomes gradually more strenuous, a condition called *presbyopia*. It is caused by the loss of elasticity of the lens due to age. This inhibits the lens from changing its curvature. The correction for presbyopia is to wear glasses.

Presbyopia is a frequent cause of visual discomfort while doing close work. It is due to the increased static muscle strength which is needed to compensate for the loss of lens elasticity. This additional muscular activity might be one of the reasons for visual fatigue. It is said that no more than two-thirds of the available accommodation power should be used to maintain a comfortable degree of focus.

Speed and accuracy of accommodation

The level of illumination is a critical factor in accommodation. When the lighting is poor the far point moves nearer and the near point recedes, while both speed and precision of accommodation are reduced as well as luminance contrast and sharpness of printed texts; the sharper the object or the character stands out against its background the quicker and more precise the accommodation.

The speed as well as the precision of accommodation decrease with age. According to Krueger and Hessen [153] these two functions show a marked decrease from about the age of 40.

17.3. The aperture of the pupil

The 'diaphragm' of the eye

Two different muscles control pupil aperture: one constricting and the other widening the pupil size. This part of the eye is called the *iris*. Its function can be compared to that of the diaphragm in a camera which is used to avoid under- or over-exposure. The pupil aperture is under reflex control to adapt the amount of light to the needs of the retina. When light levels increase the iris constricts and the pupil size is reduced. When light levels decrease the iris opens, making the pupil larger. For any given lighting condition the pupil is in a resting position as soon as the pupil size has stabilized. Even in this state, however, the pupil is in constant motion, much like the accommodated lens of the eye.

Speed of pupil reaction

The adjustment of the aperture of the pupil takes a measurable time which may vary from a few tenths of a second to several seconds. Fry and King [77] demonstrated that when stimuli producing a significant change in pupil size are presented at a slightly higher rate, about 3 Hz, than the pupil can respond to, the pupil reaction is dampened and discomfort is produced. In fact, if the level of lighting changes frequently and strongly, there is a danger of over-exposure of the retina, since the reaction time of the pupil is comparatively slow.

Brightness and pupil size

Pupil size reflects to a large extent the brightness of the visual field. It seems that the central vision is of greater importance for the regulation of the pupil size than the outer areas of the retina. During daylight the aperture may have a diameter of 3−5 mm, increasing at night to more than 8 mm. Experiments have shown that for extreme conditions pupil diameter may range from 2·25 to 6·3 mm.

Other regulating factors

The aperture of the pupil is also affected by two other factors:

1. The pupil contracts when near objects are focused and opens when the lens is relaxed.
2. The pupil reacts to emotional states, dilating under strong emotions such as alarm, joy, pain or intense mental concentration. The pupil narrows with fatigue and sleepiness.

Under normal conditions, however, the general level of lighting is the dominant regulating factor of pupil size.

Pupil size and acuity

When the pupil becomes smaller, the refractive errors of the lens are reduced and this improves visual acuity. One of the reasons that higher levels of lighting increase visual acuity is the effect of light on reducing the pupil size. Here too, it is possible to make a comparison with the camera: a small aperture of the diaphragm will increase the depth of field and generate a sharper image.

17.4. The adaptation of the retina

If we look at the headlights of a car at night we are dazzled, but the same headlights do not dazzle in daylight. If we walk from daylight into a darkened cinema where the film has already started we can see very little at first, but after about 5–10 minutes the surroundings of the room gradually become visible. These are everyday examples of how the sensitivity of the retina is continuously adapted to the prevailing light conditions. In fact, this sensitivity is many times higher in darkness than in daylight.

The process is called adaptation and comes about through photochemical and nervous regulation of the retina. Thanks to this facility we can see almost as well in moonlight as in the brightest sunlight, even though the level of illumination has increased more than 100 000 times.

Adaptation to darkness

Adaptation to darkness or to brightness takes a comparatively long time. Darkness adaptation is very quick in the first five minutes, becoming progressively slower afterwards; 80% adaptation takes about 25 minutes and full adaptation takes as much as one hour. Hence sufficient time must always be allowed for darkness adaptation, at least 25–30 minutes for good night vision.

Adaptation to light

Light adaptation is much quicker than darkness adaptation. The sensitivity of the retina can be reduced by several powers of ten in a few tenths of a second. Yet light adaptation, too, continues over a measurable time of the order of several minutes.

The abrupt reduction in sensitivity during light adaptation involves the entire retina. Whenever the image of a bright surface (a window, a light source or a bright reflection) falls onto any part of the retina, sensitivity is reduced all over, including the fovea centralis. This phenomenon, most impor-

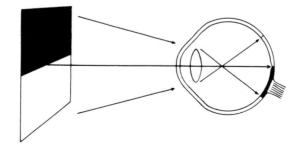

A bright area reduces the sensitivity of the whole retina

Figure 144. Effects of bright and dark surfaces on the retina.
A bright patch reduces the sensitivity of the entire retina, and thereby reduces the visual acuity in the fovea. This form of disturbance is called relative glare.

tant for precision work or for reading tasks, is illustrated in Figure 144.

Partial adaptation

In other words: if the visual field contains a dark or a bright area, adaptation will occur in the corresponding part of the retina. This adaptation appears in one part of the retina and is called local or *partial adaptation*. But, as mentioned before, this partial adaptation spreads over the whole retina, including the fovea. Such partial adaptation will therefore change the sensitivity of the retina and affect vision.

Furthermore, adaptation of one eye has some corresponding effect on the other, a fact that may be significant at workplaces where only one eye is employed.

Ergonomic principles

Two general ergonomic principles can be deduced from this knowledge:

1. *To avoid dazzle effects, all important surfaces within the visual field should be of the same order of brightness.*
2. *The general level of illumination should not fluctuate rapidly because pupil reactions as well as retinal adaptation is a relatively slow process.*

Glare

Physiologically speaking, *glare is a gross overloading of the adaptation processes of the eye, brought about by overexposure of the retina to light.* Three types of glare may be distinguished:

1. *Relative glare*, caused by excessive brightness contrasts between different parts of the visual field.
2. *Absolute glare*, when a source of light is so bright (e.g., the sun) that the eye cannot possibly adapt to it.
3. *Adaptive glare*, a temporary effect during the period of light adaptation; e.g., on coming out of a dark room into bright daylight outside. This phenomenon is also called 'transient adaptation'.

Practical hints

In this context the following hints are important for the layout of workplaces:

1. The effect of relative glare is greater the nearer the source of dazzle is to the optical axes, and the larger its area.
2. A bright light above the line of sight is less dazzling than one below or to either side.
3. The risk of dazzle is greater in a dim room since the retina is then at its most sensitive.

17.5. Eye movements

Tremor

The eyeball has several external muscles which direct the eye to the point of interest. It continuously makes small movements which keep the retinal image in a constant slight motion. Without this continuous tremor the perceived image would fade away. This is like placing your hand lightly on a rough surface and feeling the roughness only as long as the fingers move back and forth.

In general, eye movements are very precise and fast. An eye movement of $10°$ may be accomplished in about 40 milliseconds.

Vergence movements

For good vision the so-called *movements of convergence and divergence* are of special importance. Binocular vision requires the optical axes of the two eyes to meet in the object being looked at, so that the image falls on the corresponding part of the retina in each eye. When viewing an object relatively near, the visual axes are turned slightly inwards in order to intersect at the distance of the object being viewed.

If the gaze is shifted to a second object, further away, the angles of the two eyes must be opened until the optical axes again cross through the object. This movement is brought about by activity of the outer eye muscles; it is a very delicate adjustment upon which distance perception depends. This specific sensitivity is gradually developed in infancy until we learn by experience to estimate distance mainly from the angular convergence of our two eyes. In monocular vision distances must be guessed from the apparent size of objects, from foreshortening by perspective and from other visual cues.

The incredible number of eye movements

The number of eye movements required when reading a book may be as many as 10 000 co-ordinated eye movements per hour [126]. Walking over a rocky track in the mountains demands even more from the eye muscles. When the head is in motion, as in walking, the external eye muscles are in constant activity to adjust the position of the eyes in order to maintain steady fixation points. That is why objects viewed by an observer, even when walking or sitting in a car, appear stable.

If the co-ordination of external eye muscles is disturbed the phenomenon of double images will appear. This can be easily demonstrated by slightly touching one eyeball with a finger. In case of excessive fatigue transitory double images can cause annoying sensations.

17.6. Visual capacities

The various functions of the eye are not usually pushed to the limits of their performance in everyday life, but it may sometimes occur in industry or under modern traffic conditions. Furthermore, visual performances are often used in laboratory experiments to evaluate the effects of various variables such as lighting or other viewing conditions. The most important visual capacities are:

Visual acuity
Contrast sensitivity
Speed of perception

Visual acuity

Visual acuity is the ability to perceive two lines or points with minimal intervals as distinct, or to apprehend form and shape of signs and discern the finest details of an object. By and large, visual acuity is the resolving capacity of the eye. The ability to resolve a one minute of arc wide separation between two signs is often considered as 'normal' acuity. In this case the minimum distance between two points in the image on the retina is 5×10^{-6} m. However, under adequate lighting conditions a person with good vision should be able to resolve an interval of about half that size.

Influences on visual acuity

Visual acuity is related to illumination and to the nature of observed objects or signs as follows:

1. Visual acuity increases with the level of illumination, reaching a maximum at illumination levels above 1000 lx.
2. Visual acuity increases with the contrast between the test symbol and its immediate background, also with the sharpness of signs or characters.
3. Visual acuity is greater for dark symbols on a bright background than for the reverse. (Bright background decreases the pupil size and reduces refractive errors.)
4. Visual acuity decreases with age, which is shown in Figure 145.

Contrast sensitivity

Sensitivity to contrast is the ability of the eye to perceive the smallest difference in luminance, and thus to appreciate niceties of shading and the slightest nuances of brightness, all of which may be decisive for the perception of shape and form.

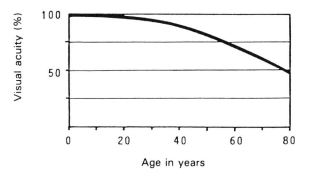

Figure 145. **Decrease of visual acuity with age.**
According to Krueger and Müller-Limmroth [154].

Contrast sensitivity is more important in everyday life than visual acuity, and this also applies for many jobs of inspection and product control. In order to measure contrast sensitivity a procedure is used in which subjects compare the luminance of a standardized target with its surroundings.

Influences on contrast sensitivity

Contrast sensitivity is subject to the following rules:

1. It is greater for large areas than for small ones.
2. It is greater when boundaries are sharp and decreases when the change is gradual or indefinite.
3. It increases with the surrounding luminance and is greatest within the range of 70 cd/m^2 and more than 1000 cd/m^2 [126].†
4. Within the mentioned range the just-perceived contrast corresponds to about 2% of the surrounding luminance, i.e. the background must be at least 2% brighter or darker than the target.
5. It is greater when the outer parts of the visual field are darker than the centre, and weaker in the reverse contrast.

Figure 146 shows results of experiments carried out as early as 1937 by Luckiesh and Moss [169]. It appears that raising the illumination level from approximately 10 lx to 1000 lx increases visual acuity from 100 to 170% and contrast sen-

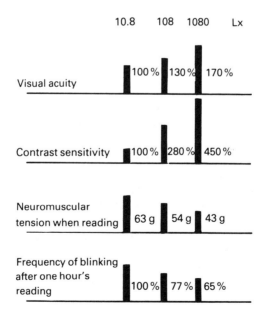

Figure 146. **Effect of light intensity on visual acuity, contrast sensitivity, nervous strain and frequency of blinking the eyelids.** *After Luckiesh and Moss* [169].

---

† For a definition of luminance and illumination see p. 251

sitivity up to 450%. At the same time the investigators recorded a decrease in muscular tension (measured from the continuous pressure of a finger on a key) and in the rate of blinking the eyelids. This was interpreted as a reduction in nervous tension as a result of better lighting.

Speed of perception

The speed of perception is defined as the time interval that elapses between the appearance of a visual signal and its conscious perception in the brain. Speed of perception is commonly measured by the technique of tachistoscopy. In this procedure a series of words is presented to the test subjects for a short time. The minimum display time required for correct perception is measured and used as parameter. Speed of perception measured by such a procedure is of course mainly a function of neural and mental mechanisms in the brain.

Speed of perception increases with improved lighting as well as with higher luminance contrast between an object (or sign) and its surroundings. That means that lighting, visual acuity, contrast sensitivity and speed of perception are closely connected with each other.

Speed of perception can be vital in transport. We need only think of an airliner flying at the speed of sound, and how much can happen during a perception time of $0 \cdot 2$ s, a common figure. But speed of perception is also an important factor in reading.

17.7. Physiology of reading

Saccades

There is a distinction between reading, as a taking in of information, and search, which involves the locating of needed information. In both activities the eyes move along a line in quick jumps rather than smoothly. These jumps are called *saccades*. They are so fast that no useful information can be picked up during their occurrence. Between the jumps the eyes are steady and fix a certain small surface which is projected. Only in the fovea and in the adjacent area, the parafovea, is detailed vision sufficiently accurate for the recognition of normal print.

Three forms of reading saccades are of importance: the *rightward reading saccades*, the *correction saccades* and the *leftward line saccades*.

The *rightward reading saccades* along a line cover in each jump an area of about 8 ± 4 letters. Occasionally small leftward saccades may occur, the so-called *correction saccades*. The *line saccades* start just before the end of a line is reached and jump to the beginning of the next line.

Bouma [25] did a thorough study on eye saccades and eye

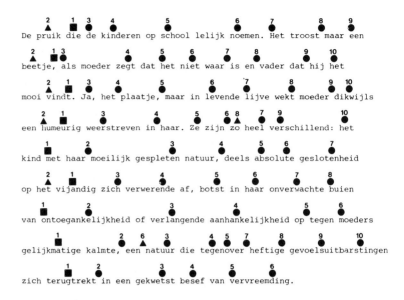

Figure 147. Saccades and fixations of eyes in the silent reading of a Dutch text.
Three different types of saccades are indicated: reading saccades (circles), correction saccades (triangles) and line saccades (squares). Numbers indicate the order of fixations within each line. According to Bouma [25].

fixations of reading subjects. Figure 147 shows the succession of eye saccades and eye fixations of a subject reading a Dutch text. All types of saccades may be different for different texts and different subjects.

Character recognition

The eye pauses between saccades last mostly between 120 and 300 ms (Bouma [25]). During these pauses characters are recognized in foveal and parafoveal vision. For rapid and good recognition it is important that characters are *acceptable*, *identifiable* and *distinctive*.

Acceptability is the degree to which characters correspond to the 'internal model' a reader has of them. This is the fundamental process of reading.

Identifiability requires clear letter details which must be designed clearly.

Distinctiveness means that each character has such a specific design that no confusion can occur. The extension of descending letters (such as p and q) and of ascending letters (b or d) can be important for good distinctiveness.

Visual reading field

During the eye pause the fovea and the adjacent area pick up visual information from a rather small surface, the so-called *visual reading field*. To code numbers without much redundancy only a few symbols can be picked up in a single glance. For words the visual reading field may be larger, because sufficient word knowledge renders the recognition of the full word possible at the sight of merely a few letters.

When reading a text the eyes make about four fixations per second. In well-printed texts the visual reading field can easily be as wide as 20 letters, about 8 to the left of fixation and 12 to the right. The visual reading fields overlap, that is to say that words within the visual reading field may appear at least twice.

According to Dubois-Poulsen [60] the following time fractions are about normal:

Gaze fixation between saccades:	$0 \cdot 07 - 0 \cdot 3$ s
Rightward reading saccades:	$0 \cdot 03$ s
Line saccades:	$0 \cdot 12$ s

Line saccades

Correct line saccades require sufficiently large distances. The lines above and below the reading line will interfere with parafoveal word recognition unless line distances are sufficiently wide. If they are too narrow the visual reading field becomes restricted so that less information can be picked up in a single eye pause. *Thus a wide visual reading field calls for sufficient interline distance.* According to Bouma [25] the visual reading area around fixation, which is free from interferences by the two adjacent lines of print, decreases by shortening the interline distance. If the reading field covers 15 letters the interline distance must be equal to about five times the height of lower case characters; if the reading field is restricted to seven letters the interline distance must still be equal to two lower case characters. The same author recommends a minimum admissible interline distance of about 1/30 of the line length (in this book it is 1/24). As a consequence, interline distance should increase with line length. For VDTs it seems advantageous to use screens vertically oriented, since such a screen design would require shorter lines and smaller interline distances.

Contrast and colour

According to Timmers [259] parafoveal word recognition is critically dependent on character contrast. The lower the contrast, the narrower is the visual reading field and the lower, therefore, the readability. Similar effects were observed on VDTs with coloured letters. Engel [68] showed that coloured letters and digits can only be read when quite close to the fixation, although colour itself may well be discernible far away from the fixation. *This indicates that colour is a useful aid for visual search but actual reading takes place in a restricted visual reading field.*

If a reader is familiar with the significance of colours, then colours will help to locate the required information quickly, but the recognition of a word or symbol itself depends on the legibility of characters and not on their colour.

17.8. Visual strain

Excessive eye strain can have two main effects: tiring the eyes, and adding to general fatigue.

Visual fatigue

Visual fatigue comprises all those symptoms that arise after excessive stress on any of the functions of the eye. Among the most important of these are straining the ciliary muscle of accommodation by looking too closely at very small objects, and the effects of strong local contrasts on the retina. Visual fatigue manifests itself as:

1. Painful irritation (burning) accompanied by lachrymation, reddening of the eye and conjunctivitis.
2. Double vision.
3. Headaches.
4. Reduced powers of accommodation and convergence.
5. Reduced visual acuity, sensitivity to contrast, and speed of perception.

These comparatively severe symptoms are brought about in particular by strenuous fine work, by reading poorly printed texts, by inadequate lighting, by exposure to flickering light or by optical aberrations such as hypermetropia (long sight). Elderly people are of course more prone to visual fatigue.

Obviously all types of visual work contribute to the general fatigue discussed earlier, since every job that calls for more rapid and precise eye movements will make heavier demands on perception, concentration and motor control of the hands. So whenever the eyes are overstressed for long periods the symptoms of eye strain (sore eyes and headaches) will be added to those of general fatigue.

The effects of visual fatigue on a person's occupation may include:

1. *Loss of productivity.*
2. *Lowering of quality.*
3. *More mistakes.*
4. *Increased accident rate.*
5. *Visual complaints.*

Accident rate

In a report by the American National Safety Council, the experts reckoned that bad lighting was the cause of 5% of all industrial accidents, and, together with the optical fatigue it engendered, contributed to as many as 20% of them.

The experience of an American heavy industry (Allis Chalmers) may be mentioned as an example [78]. After the level of illumination over an assembly line had been increased to 200 lx, there was a fall of 32% in the accident rate. As a further step the walls and ceilings were painted in light colours, to reduce contrast and provide a more uniform illumination, and the accident rate fell by another 16·5%. Similar surveys in England and France showed drastic reductions in accident rates when the lighting conditions were improved, especially in shipyards, foundries, large assembly lines and engineering shops.

Lighting and productivity

There are also many reports of increased productivity after the lighting was improved. These increases are partly a direct effect (through more rapid visual assessment of the work), and partly indirect, through the reduction in fatigue. McCormick and Sanders [180] give a table summarizing the results of 15 industrial studies, all of which showed increases in output, ranging from 4 to 35%, after increasing the level of illumination. The original level had been very low, however, less than 100 lx. McCormick himself [180] had reservations, because of the existence of other, uncontrollable factors that were always present in such situations. Yet in spite of this valid criticism, there is no doubt that the increase was partly due to the previous inadequate lighting.

An interesting survey in an American cotton-spinning factory showed a stepwise improvement in productivity when the general illumination level was increased. When the illumination was raised from 170 to 340 lx, the production rose by about 5%, while simultaneously the amount of rejected product was sharply reduced. As a result, the total costs fell by 24·5%. This result encouraged the management to increase the illumination still further up to 750 lx, whereupon production rose to 10·5% above the original level, and the reduction in wastage brought costs down by almost 40%.

Similar results were obtained in England, France, Germany and other countries, often showing increases in productivity, reduction in rejected products, and fewer accidents as the level of illumination was increased.

Visual strain of VDT operators

The expansion of VDTs in the last decade has been accompanied by complaints by many VDT operators about visual strain. Systematic research has been carried out in various countries. For detailed results the reader is referred to the book *Ergonomics in Computerized Offices* [85]. Most studies disclosed an increased incidence of visual discomfort together with the above-mentioned symptoms of visual fatigue. A few studies, however, did not confirm these results since the frequency of complaints among VDT operators did not significantly exceed that of control groups. The controversial results might to some extent be explained by the choice of control groups. Complaints of visual discomfort might be frequent in control groups engaged in strenuous office work, but occur more seldom among control groups occupied in traditional office work.

Correlates of visual discomfort	Some studies revealed significant relationships between the photometric characteristics of VDTs and symptoms of visual discomfort. Thus, screen flicker, excessive luminance contrast ratio between screen and environment, reflected glare on the screen and poor readability were related to an increased incidence of visual complaints. These findings lead to the assumption that sharpness, luminance contrasts, stability, character flicker, screen reflections and the geometric design of characters might decrease the legibility and produce occasional visual fatigue. Bräuninger *et al.* [26] measured these photometric characteristics for a great number of VDT makes and models and found a lot of insufficient photometric parameters.

We shall restrict ourselves here to summarizing the recommendations deduced from these studies.

Recommended luminances and contrasts	Displays with bright characters should neither exhibit a too dark nor a too bright screen background. A luminance contrast ratio between background and characters of 1:6 is already sufficient for good readability. Given a character luminance of $40-50$ cd/m^2, a background luminance of $6-8$ cd/m^2 would be appropriate.

Degree of oscillation of character luminances	The display must be free of flicker for all operators. As a general rule it can be recommended to lower the degree of character oscillation to a level comparable to figures shown by phase-shifted fluorescent tubes (see Figure 148). Refresh rates of $80-100$ Hz with a phosphor decay time of approximately 10 ms for the 10% luminance level are recommended.

Sharpness...	The characters should show sharp edges; no blurred border zone should be perceived. If the blurred border zone is less than $0\cdot3$ mm, characters appear to have sharp borders.

Poor sharpness is often due to an insufficient focusing device, a too high adjusted character luminance or unsuitable antireflective devices.

...and stability	The electronic control of the electron beam must ensure good character stability. Neither drifts nor jitter should be perceived by the operators.

Reflections on screen surfaces	All antireflective technologies available on the market today have serious drawbacks: some are associated with a decrease of sharpness and an excessive dark screen background, others are easily soiled. If efficiency is weighed against drawbacks, the quarter-wave coatings and the etching-roughening procedures are to be preferred. Reflected glare on the screen surface should be reduced by $5-10$ times.

Size of
characters and
face

The range for appropriate character sizes on VDTs is 16−25 minutes of visual angle. This means that 3 mm is a suitable height for characters at a viewing distance of 500 mm and 4·3 mm at 700 mm. The following sizes are recommended:

Height of capital letters	3−4·3 mm
Width of characters	75% of height
Distance between characters	25% of height
Space between lines	100−150% of height

The spaces between dots should not be visible. Thus, a dot matrix of 7×9 offers better legibility than one of 5×5.

All symbols should have shapes that are easily distinguished from each other and acceptable as reasonable representations of the symbols concerned.

Dark versus
bright characters

VDTs with dark characters and a bright screen background offer the following advantages: reading conditions similar to printed texts, low contrast ratios to the visual environment and less disturbing reflections on the screen. The drawbacks, at present, are an increased risk of flicker and a small stroke width. Thus a refresh rate of 90 Hz, a phosphor with a decay time of approximately 10 ms to reach the 10% luminance level and a stroke width of about 0·38 mm are recommended.

18. Ergonomic principles of lighting

18.1. Light measurement and light sources

In order to understand what follows later it is worth defining two of the many terms employed in the study of lighting, *illumination and luminance*.

Illumination

Illumination is the measure of the stream of light falling on a surface. The light may come from the sun, lamps in a room or any other bright source. The unit of measurement is the *lux*, defined as

1 lux (lx) = 1 lumen (lm) per square metre, the lumen being the unit of luminous flux.

A formerly used unit in the English speaking world was the *footcandle* (ft c). 1 lux is approximately $0 \cdot 1$ footcandle $(0 \cdot 0929$ ft c).

The human eye responds to a very wide range of illumination levels, from a few lx in a darkened room to approximately 100 000 lx outside under the midday sun. Illumination levels in the open vary between 2000 and 100 000 lx during the day, whereas at night artificial light of 50—500 lx is normal.

Luminance

Luminance is the measure of the brightness of a surface; the perception of brightness of a surface is proportional to its luminance. Therefore, *luminance is a measure of light coming from a surface.* Since it is a function of the light that is emitted or reflected from the surface of walls, furniture and other objects, it is greatly affected by the reflective power of the surface. The luminance of lamps on the other hand is an exact measure of the light they emit. Bright characters on a dark background in VDTs are also emitting light which can be characterized by its luminance.

Candela/m²

In the metric system *luminance is measured in units of candela per m² (cd/m²).* In the past the standard reference for measuring luminance was actually a wax candle of certain specifications. Today the standard is much more precise, but the terminology stems from the earlier concept.

Millilambert and
footlambert

In the English speaking world the terms *millilambert* (mL) and *footlambert* (ft L) are still used to measure luminance. One millilambert (mL) is the amount of light emitted from a surface at the rate of $0 \cdot 001$ lumen/cm². A footlambert (ft L) is the amount of brightness of an ideally reflecting surface illuminated by one footcandle.

The following equations apply:

1 cd/m² $= 0 \cdot 292$ footlambert (ft L)
1 footlambert (ft L) $=$ ca. $3 \cdot 5$ cd/m²
1 millilambert $= 3 \cdot 183$ cd/m²
1 footlambert $= 1 \cdot 076$ millilamberts

Fortunately, the cd/m² has gradually become the most frequently used unit to define the luminance of surfaces.

A few examples will illustrate the approximate luminance of some common sources of light in an office with an illumination of 300 lx:

Fluorescent lamp (65 watt)	10 000 cd/m²
Window surface	1000 − 4000 cd/m²
White paper lying on a table	70 − 80 cd/m²
Table surface	40 − 60 cd/m²
Bright enclosure of a VDT	70 cd/m²
Dark enclosure of a VDT	4 cd/m²
Screen background	5 − 15 cd/m²

Reflectance

If the luminances of various surfaces are compared they can also be expressed as *reflectance*, which is the ratio between incident and reflected light. *It is usually expressed as the percentage of reflected to incident light.* The luminance in cd/m² and the illumination in lx are related as follows:

$$\text{Reflectance } (\%) = \frac{\text{cd/m}^2}{\pi \text{ lx}} \times 100$$

A simple example is this: if a bright table surface has a reflectance of 70% and the incident light has an illumination figure of 400 lx, the luminance of the table will then be 70% of $400/\pi = 89$ cd/m².

Direct and
indirect lighting

Among the various lighting systems in offices one can distinguish between direct and indirect lighting.

Directional lighting sends about 90% of its light towards targets in the form of a cone of light. These light sources cast hard shadows with sharp contrasts between light and shadow. Excessive contrast tends to produce relative glare. Directional lighting systems can be recommended in offices as working lights only where the general illumination is high enough to reduce this contrast. At VDT workstations such lighting is used when the general illumination is insufficient for reading source documents with poor legibility.

Indirect lighting throws 90% or more of its light onto the ceiling and walls which reflect it back into the room. This system requires the ceiling and walls to be light-coloured. Indirect lighting generates diffuse light and casts practically no shadows. In a traditional office (without business machines) it can give a high level of illumination with a low risk of glare. In offices with VDTs the bright ceilings and walls can produce reflections on the screens and cause relative glare.

A widely used system is a *combination of direct and indirect lighting*. The luminaires have a translucent shade and about 40−50% of the light radiates to the ceiling and walls while the rest is thrown directly downwards. This type of lighting casts only moderate shadows with soft edges. The whole room, including furniture and shelves placed at the walls, is evenly lighted.

Opalescent globes and similar free radiants, give out light equally in all directions, and throw slight to moderate shadows. Because they are bright, they often cause glare, and so they should not be used in living rooms or work rooms. They are suitable for store-rooms, corridors, entrance halls, vestibules, lavatories, etc.

Light sources

Light sources are mainly of two kinds: *electric filament lamps* and *fluorescent tubes*.

Filament lamps

The light of filament lamps is relatively rich in red and yellow rays. When used above a workplace they emit heat. Lamp-shades can reach temperatures of $60°C$ and more and can cause discomfort and headaches. On the other hand, their warm glow creates a pleasant atmosphere.

Fluorescent tubes

Fluorescent lighting is produced by passing electricity through a gas (argon or neon) or through mercury vapour. This procedure converts electricity into light much more efficiently than a heated filament. The inside of the tube is covered with a fluorescent substance which converts the ultraviolet rays of the discharge into visible light, the colour of which can be controlled by the chemical composition of the fluorescent material. Fluorescent tubes have a series of advantages:

Advantages

High output of light and long life.
Low luminance, when adequately shielded.
Ability to match the light to daylight or at least to a pleasant and slightly coloured light.

Drawbacks

But they also have serious drawbacks. Since they operate from alternating current, fluorescent tubes produce a flicker-ing light, at a frequency of 100 Hz in Europe and 120Hz in the USA. *This is above the normal flicker-fusion frequency, the so-called critical fusion frequency (CFF) of the human eye*, but it can become noticeable as a stroboscopic effect on moving objects. Furthermore, old or defective tubes develop a slow visible flicker.

Visible flicker

Visible flicker has adverse effects on the eye mainly because of the repetitive over-exposure of the retina. Flickering light is extremely annoying and causes visual discomfort.

It is generally assumed that the luminance oscillation of fluorescent tubes with a rate of 100 Hz and more is above the critical fusion frequency and therefore cannot affect the eye. Several studies, however, indicate that the exposure to single fluorescent tubes may have adverse effects on human subjects. Our own studies [88] have shown that working lamps with single fluorescent tubes can increase visual fatigue and measurably reduce the performance of fine assembling work. Wiebelitz and Schmitz [273] observed a decrease in the pupil reaction to light when subjects were reading under fluorescent light with flicker frequencies between 25 and 100 Hz. Raising the flicker frequency to 200 Hz or more removed this effect. Recent experiments on cats carried out by Eysel and Burandt [69] revealed that the visual system in the brain responds distinctly to the temporal information present in light from fluorescent tubes driven by 50 or 60 Hz alternating current. This study confirmed earlier findings showing that the critical fusion frequency in the optic tract is above 100 Hz. But these findings cannot be considered to be a definite proof for adverse effects of fluorescent light with invisible flicker of 100 or 120 Hz.

Early experiences with fluorescent light

A general experience of rather anecdotal character must be mentioned here. When fluorescent lighting was first introduced on a large scale in European offices, a series of complaints about irritated eyes and eye strain were reported. On the assumption that the oscillating character of fluorescent light was the cause of visual discomfort, phase-shifted equipment was developed which produced an almost constant light. Complaints seem to have stopped in offices where phase-shifted fluorescent tubes have been installed.

Subharmonic 50 Hz oscillations?

A study by Collins [47] in 1957 revealed another interesting aspect of fluorescent tubes. With a number of different models of fluorescent tubes he recorded small 50 cycle per second fluctuations superimposed on the main 100 cycle one. These subharmonic 50 Hz oscillations come from a partial rectifying action in the discharge due to asymmetrical emissions by the electrodes. Ten per cent of both old and new tubes were observed to have this effect. Small amounts of subharmonics were found to be perceptible by subjects and the author assumed that such tubes are sufficiently common to account for the complaints which had arisen with fluorescent lighting.

Nevertheless the question of what degree of oscillation in fluorescent tubes is acceptable remains unsettled.

Phase-shifted fluorescent tubes

In Europe it was generally resolved that *offices should never be lit with single fluorescent tubes but always with two or more phase-shifted tubes inside one luminaire.*

Figure 148 shows recordings of the luminance oscillation of fluorescent tubes. It illustrates the effects of phase-shifted equipment which generates an almost constant luminance.

It is obvious that appropriate equipment will avoid the disadvantages of fluorescent tubes, so that their undisputed advantages can be fully utilized.

3 fluorescent tubes:

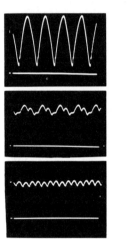

Without phase difference
UF = 0.157

In 2 phases
UF = 0.77

In 3 phases
UF = 0.89

Figure 148. Effects of different arrangements of fluorescent tubes on the uniformity of the light.
The curves express the invisible flicker, as registered by a photoelectric cell, with the horizontal lines marking the zero level of light intensity. UF (uniformity figure) = min/max light intensity.

18.2. Physiological requirements of artificial lighting

For visual comfort and good optical performance, the following conditions should be met:

Suitable level of illumination.
Spatial balance of surface luminances.
Temporal uniformity of lighting.
Avoidance of glare with appropriate lights.

The physiological requirements under these four headings are just as valid for artificial light as for natural daylight, but since the practical problems are somewhat different, the requirements for artificial light will be considered first.

Illumination levels

60 years ago illumination levels of 50−100 lx were generally recommended for workshops and offices. Since then the figures have increased steadily and today levels between 500 and 2000 lx are quite common. The general attitude towards lighting has been 'the more the better'. But this does not necessarily hold true for offices.

Between 1960 and 1968 Blackwell [19] used an optical test which took into account the ambient lighting, the size of the object, the contrast and the speed of perception. The results of these studies are embodied in the US standards of the IES [127].

If the current recommendations are compared, it will be seen that the standards of the American Illuminating Engineering Society (IES) prescribe significantly higher levels of lighting than are embodied in the European guidelines. A few of these proposals are given in Table 31.

Table 31. Comparison between German (DIN) and US (IES) standards for intensity of lighting, each in lx.

	DIN [57]	IES [127]
Rough assembly work	250	320
Precise assembly work	1000	5400
Very delicate assembly work	1500	10800
Rough work on toolmaking machine	250	540
Fine work on toolmaking machine	500	5400
Very precise work on toolmaking machine	1000	10800
Technical drawing	1000	2200
Book-keeping; office work	500	1600

Drawbacks of too high illumination levels

A study in 15 open plan offices [198] has shown that a very high level of illumination is often unsuitable in practice. Levels above 1000 lx increase the risk of troublesome reflections, deep shadows and excessive contrasts. In the above-mentioned study 23% of 519 employees reported that they were disturbed by either reflections or glare.

Another interesting observation in the same study was the significantly higher incidence of eye troubles in offices with illumination levels above 1000 lx.

All employees preferred illumination levels between 400 and 850 lx. Obviously it would be going too far to interpret these results to the effect that there is a direct causal relationship between illumination level and eye troubles, but there are good reasons to believe that brightly lit open plan offices more often create reflections, deep shadows and relative glare, and that these possibly contribute to the eye troubles recorded. In other words: illumination levels of over 1000 lx do not directly

cause visual discomfort, but their consequences on the balance
of luminances are the main source of trouble. These results
conflict with several studies carried out in well-lit test rooms
with preferred illumination levels of 1000–4000 lx. The carefully designed brightness of the surroundings may account for
the contradictory results.

As far as current results go, the values given in Table 32
may be recommended as a basis for comparison of work-rooms
for different purposes.

If a strong light is necessary this is best achieved by the use
of working lights, but these should always be used in conjunc-
tion with a good general illumination, to avoid creating too
much contrast. Guidelines might be as follows:

Working light(s)	General illumination
500 lx	150 lx
1000 lx	300 lx

Specifications for lighting levels can be no more than general
guidelines, and other circumstances must be taken into
account in any particular situation. For example:

1. The reflectivity (colour and material) of the working materials and of the surroundings.
2. The extent of the difference from natural lighting.
3. Whether it is necessary to use artificial lighting during the
   daytime.
4. The age of the people concerned.

Table 32. **Examples of suitable lighting levels in work-rooms.**

Type of work	Examples	Recommended lighting (lx)
General	Storeroom	80–170
Moderately precise	Packing; despatch; works laboratory; simple assembly; winding thick wire on to spools; work on carpenter's bench; turning; boring; milling; locksmith's work	200–250 250–300
Fine work	Reading; writing; book-keeping; laboratory technician; assembly of fine equipment; winding fine wire; woodworking by machine; fine work on toolmaking gig	 500–700
Very fine to precision work	Technical drawing; colour proofing; adjusting and testing electrical equipment; assembling delicate electronics; watchmaking; invisible mending	 1000–2000

Age

Blackwell [19] studied the effect of age with his test procedure. If the degree of contrast necessary to satisfy people in the age-group 20−25 was taken as unity (1), then for older people this must be multiplied by the following factors:

40 year olds: 1·17
50 year olds: 1·58
65 year olds: 2·66

Spatial balance of surface luminances

The distribution of luminances of large surfaces in the visual environment is of crucial importance for both visual comfort and visibility. In general, the higher the ratio of change or difference in luminance levels, the greater the loss of visibility.

How to express luminance contrast

Although there are many ways to define relative luminances, the most common one is simply to specify the ratio of two luminances. Many authors, however, prefer the definition adopted by the International Lighting Commission (CIE) according to the following formula:

$$C = \frac{L_0 - L_B}{L_B}$$

where C=contrast; L_0=luminance of the target; and L_B=luminance of the background.

Sharp contrasts, e.g., relative or transient glare

It is generally agreed that sharp luminance contrasts between large surfaces located in the visual environment reduce visual comfort and visibility, but the degree of acceptable contrast ratios is contested. Tolerable luminance ratios depend upon specific circumstances. Many factors are involved, such as size of the source of glare, its distance from the viewer's line of sight and the intensity of the general illumination in the room. Furthermore, results of experiments will also depend on whether quantitative visual performance or subjective visual discomfort is measured.

The early study by Guth

Among the earlier studies, the results of Guth [98] shall be mentioned here. He observed in subjects a decrease in contrast

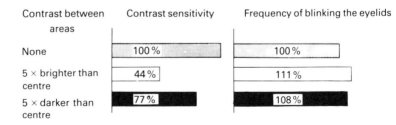

Figure 149. Physiological effects of contrasting areas in the middle of the visual field.
The contrast is measured between the 15% of the visual field in the centre and the area immediately adjacent to this. After Guth [98].

sensitivity and an increase in the eye blinking rate when the centre area of the visual field was five times brighter or darker than the adjacent area. Figure 149 shows the results of this study.

According to these experiments relative contrast ratios of 1:5 in the middle of the visual field significantly impair the efficiency of the eye as well as visual comfort. If the adjacent areas are brighter than the centre area the disturbances seem to be more strongly felt than vice versa.

General rules   The following general rules are widely accepted:

1. All the objects and major surfaces in the visual field should as far as possible be equally bright.
2. Surfaces in the middle of the visual field should not have a brightness-contrast of more than 3:1 (see Figure 150).
3. Contrast between the middle field and the edge of the visual field, or its vicinity, should not exceed 10:1 (see Figure 150).
4. The working field should be brightest in the middle, and darker towards the edges.
5. Excessive contrast is more troublesome if it occurs at the sides of and below the visual field, than at the top of the field.
6. Light sources should not contrast with their background by more than 20:1.
7. The maximum permissible range of brightness within the entire room is 40:1.

In everyday practice these guidelines are often neglected. As always one comes across contrasting elements which

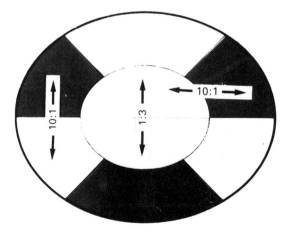

Figure 150. Acceptable contrasts between the brightness of various parts of the visual field.
Within the middle field, 3:1; within the outer field 10:1; between middle and outer field 10:1

should be avoided in the visual environment, such as:

Bright walls.

Dazzling white walls contrasting with dark floorings, dark furniture or black office machines.

Reflecting tabletops.

Black typewriters on bright underlays.

Polished machine parts.

The choice of colour and material is of great importance in the design of walls, furniture and larger objects in a room, because of their varying reflectance. The following reflectances are recommended:

Ceiling:	80–90%
Walls:	40–60%
Furniture:	25–45%
Machines and equipment:	30–50%
Flooring:	20–40%.

Windows must always be equipped either with adjustable venetian blinds or with translucent curtains, so that excessive contrast can be avoided on sunny days. Workplaces should be at right-angles to the window, as shown in Figure 151. This applies equally to schoolrooms, meeting rooms, conference halls, libraries, etc.

Workplaces where most delicate visual work is being done are exceptions to this rule. Here the light must come from in front, so the bench is often placed across the window, and then

Figure 151. Correct and incorrect arrangement of the work place.
Left: the secretary, when reading, has a bright window surface in her visual field; the contrast between the window and other surfaces is more than the ratio 10:1.
Right: the bright window is not in the visual field of the secretary; the luminance contrasts have been decreased and the ratio corresponds to the recommendations of Figure 150.

the operator, to avoid glare, must bend his head so far forward that he is almost vertically down on his work. Hence frontal lighting is often a cause of bad posture of the neck and body.

*Designers'
imagination*

Some designers come up with very innovative and individual ideas in an attempt to design attractive furniture for offices. They visualize pitch-black office machines on a bright table or dark furniture neighbouring bright walls. Such designers don't care about ergonomic principles or balanced surface luminances.

The instructions for designers of a VDT workstation can therefore be summed up as follows: *Select colours of similar brightness for the different surfaces, renounce eye-catching effects with black and white contrasts, avoid reflecting materials and give preference to matt colours.*

The recommendation to avoid luminance contrasts exceeding a ratio of 3:1 in the middle of the visual field is generally not accepted for VDT workstations. This problem is discussed on p. 267.

*Temporal
uniformity of
lighting*

Even more disturbing than static contrast are rhythmically fluctuating bright areas in the visual field. These occur if the work requires the operator to glance alternately at a bright and a darker object; if bright and dark objects pass by on a conveyor belt; if moving parts of a machine are bright and reflective; or if a lamp itself flickers.

As we have already seen, the pupil and the retina of the eye can cope with changes in brightness only after a certain delay, so that fluctuating brightness leaves the eyes either under- or over-exposed for much of the time. Hence such lighting conditions are particularly disturbing. Physiological research has shown that if two brightness levels in the ratio 1:5 fluctuate rhythmically, visual performance is reduced as much as if the level of illumination had been lowered from 1000 lx to 30 lx.

To avoid fluctuating levels of brightness as far as possible:

1. *Cover moving machinery with an appropriate housing.*
2. *Equalize brightness and colour along the main axes of sight.*
3. *Take the precautions mentioned earlier to avoid flickering light sources.*

18.3. Appropriate arrangement of lights

*Avoid glare with
appropriate
lights*

Inadequate lights or lighting arrangements can be sources of glare which make viewing difficult and uncomfortable. *To avoid glare inside a room is one of the most important ergonomic considerations when designing offices.*

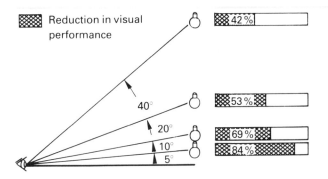

Figure 152. Effect of source of glare (dazzle) on visual performance.
The hatched blocks indicate the reduction in visual performance as a percentage of the normal performance when there is no glare present. Visual performance becomes worse, the closer the light source is to the optical axis. After Luckiesh and Moss [169].

Figure 152 sets out the results of a classical piece of research by Luckiesh and Moss [169]. Their test subjects carried out a visual task in which a light source of 100 watt was moved closer to the optical axis step by step. Visual performance was gradually impaired.

A bad example   Figure 153 shows a very unsatisfactory arrangement of lights. Opalescent globes are being used in a drawing office, where they will often come into the visual fields of the draughtsmen, as well as being reflected back from the polished floor covering. The result is very strong contrasts, far exceeding the recommended maximum of 10:1.

Figure 153. Unsuitable lighting in a drawing office.
The opalescent globes are sources of much glare. The dark floor contrasts too strongly with the white working surfaces (relative glare), and throws back strong reflections of the lamps.

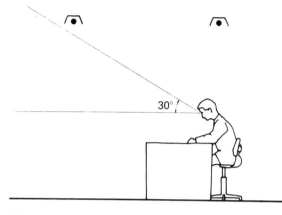

Figure 154. The angle between the horizontal and the direction from eye to overhead lamp should be more than 30°.

The following recommendations should be considered, in order to arrive at a good arrangement of lights, and appropriate overall distribution of light:

1. *No source of light should appear in the visual field of any worker during working operations.*
2. *All lights should be provided with shades or glare shields to prevent the luminace of the light source from exceeding 200 cd/m².*
3. *The line from eye to light source must make an angle of more than 30° with the horizontal* (see Figure 154). If a smaller angle cannot be avoided, e.g., in large workrooms, then the lamp must be effectively shaded.
4. *Fluorescent tubes should be aligned at right angles to the line of sight.*
5. *It is better to use more lamps, each of lower power, than a few high-powered lamps.*

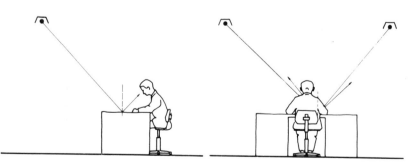

Figure 155. Left: poor placing of a single lamp so that its reflection comes into the line of sight of the operator, with the risk of direct glare.
Right: the reflections of two lamps placed to the sides are not in the line of sight, so reflected glare is avoided.

6. *To avoid glare from reflection the lines from workplace to lamp should be such that they do not coincide with any of the directions in which the operator normally needs to look.* No reflection giving a contrast greater than 10:1 should come within the visual field; see Figure 155.
7. *The use of reflective colours and materials on machines, apparatus, table tops, switch panels, etc. should be avoided.*

18.4. Lighting for fine work

Fine and delicate work

Very precise work, say from the size of normal print down to fractions of a millimetre, needs special lighting to supplement the general illumination. Examples of such work include:

1. Colour testing in chemical works, a paper factory and the textile industry.
2. Delicate assembly work; adjusting and testing electronic equipment; making watches and clocks; precision engineering.
3. Grinding, etching, polishing and engraving glass.
4. Weaving, sewing, knitting, colour printing, invisible mending and quality testing in the textile industry.

Very small objects may need to be magnified, and lenses, magnifying glasses and other optical aids should be provided.

Requirements for good vision

The following considerations are important in this context:

1. Level of illumination at the workplace.
2. Distribution of bright surfaces with the visual field.
3. Size of the objects to be handled.
4. How much light is reflected from these objects.
5. Contrast between objects and surroundings and shadows.
6. How much time is available for seeing whatever is necessary.
7. The age of the persons concerned.

As already mentioned, fine and delicate work requires illumination levels between 1000 and 10 000 lx. Such high levels of illumination are usually demanded by operators who need to concentrate on very tiny objects, or wish to throw objects into relief by creating strong contrast.

So we can deduce the general principle that work demanding high visual acuity (recognition of the tiniest shapes or objects) and contrast sensitivity (control of colour or pattern in textile, checking X-ray pictures, etc.) calls for a high level of illumination. The values given in Tables 31 and 32 can be recommended as guidelines.

Lights that are too bright can be damaging

Occasionally, however, the lighting can be too bright and harmful. Reflections from metallic surfaces may impair vision. Furthermore, fine structures in materials or surface irregularities in metal may actually be seen more easily with moderate lighting than if over-illuminated.

Contrasts in fine work

In contrast to large areas, very small objects are better lit with strong contrasts; thus dark markings or objects on a light background are easier to see than bright objects on a dark background. For this reason, when working with very small objects, it is better to have the work lit from in front rather than from the side, since then the back of the object is in deep shadow and the object stands out against the bright, reflective working surface.

Frontal lighting

Thus the incidence of the light and the casting of shadows can make a considerable difference to the recognition of objects and to the interpretation of their surface structure.

Very diffused light, without shadows, makes everything look flat and featureless, whereas lighting that casts shadows makes things more solid.

Current knowledge leads to the conclusion that for very fine work in industry, neither a completely diffused light, nor one full of deep shadows, is entirely suitable. We had experience in relation to the checking of metal parts for uneven areas and spots of rust. This visual task was easier with a lamp that was half diffused than with one that gave full, direct light. Figure 156 shows this workplace diagrammatically.

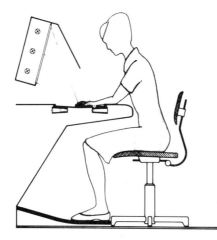

Figure 156. Diagram of lighting at a workplace where metal parts passing on a conveyor belt are inspected, and defective parts discarded.
The light comes from three fluorescent tubes, out of phase, giving a large light source, with a diffuser in front of it, and a shade to protect the eyes from direct light. The frontal lighting enables defective parts to be quickly recognized.

Different kinds and arrangements of light for very fine work should always be tested with experienced workers. Many kinds of precise work pose special visibility problems which cannot be solved by stereotyped methods.

General recommendations for fine work

In spite of this reservation we can formulate general principles that are especially valid for precise assembly work, or delicate mechanical tasks:

1. Use frontal lighting.
2. Screen the lamps from being directly visible.
3. Lamps should have diffusive screens of ribbed or frosted glass. This half diffused light is reflected less and creates less contrast.
4. The light source should emit from a large area.
5. The diffusing screens should be broad and deep so as to make the illumination of the work bench as uniform as possible.
6. Phase-shifted fluorescent tubes are preferable to filament lamps, since the latter give off more heat.

18.5. Lighting in VDT offices

Illumination levels for VDT workstations

The general recommendations for illumination levels are not valid for offices with VDT workstations. A VDT operator who is alternately looking at a dark screen and a bright source document is exposed to great luminance contrasts. We will see on the following pages that the contrast ratio between screen and source document should not exceed a figure of 1:10, which implies that the illumination level on source documents should be kept low. On the other hand, however, the reading task requires that the source document be well-illuminated. This conflicting situation calls for a compromise. Hence it is not surprising that the assessment of the optimum illumination level is a controversial matter.

Preferred illumination levels

Walking through offices with VDT workstations one can often notice that single fluorescent tubes have been removed or switched off by operators. Upon enquiry they cannot give plausible reasons for this but claim that a lower illumination level suits them better.

Some research has been done in the field of preferred lighting conditions at VDT workstations. Shahnavaz [239] carried out a field study in a Swedish telephone information centre. The operators could adjust the level of illumination on the working desk. The preferred mean illumination levels on the telephone catalogue were 322 lx during the day and 241 lx for night-shifts with similar levels on the desk and on the keyboard.

A study in Germany carried out by Benz *et al.* [15] revealed that 40% of the VDT operators preferred levels between 200 and 400 lx, whereas 45% had levels between 400 and 600 lx.

During a study of 38 CAD (computer-aided design) workstations van der Heiden [108] noticed that a number of lights had been switched off, reducing the mean illumination levels to around 120 lx.

Recommended illumination levels

It is clear that ambient office illumination should be reduced to a level compatible with the work at VDTs. Such levels are in the order of 200 lx but generally cause the office to appear dimly lit. Furthermore, 200 lx is in most cases inadequate for reading hard-copy documents.

Finally, it must be emphasized that it is not suitable to recommend merely one figure since the working conditions might differ from one job to another. For instance, Läubli *et al.* [158] observed in a field study that operators working on data-entry tasks tended towards higher illumination levels than those engaged in conversational tasks.

General experience as well as several field studies lead to the recommendations given in Table 33.

Table 33. Recommended illumination levels at VDT workstations.
The lx figures refer to measures taken on a horizontal plane.

Working conditions	Illumination level (lx)
Conversational tasks with well printed source documents	300
Conversational tasks with reduced readability of source documents	400−500
Data entry tasks	500−700

Surface luminances at VDT workstations

The surfaces in the visual field and in the visual surroundings of a VDT operator are screen, frame and enclosure of display, desk, keyboard, source documents and other elements of the immediate environment, such as walls, windows, ceiling and furniture.

The contrasts of surface luminances at VDT workstations with bright characters and dark background are often excessive, similar to the contrast between the dark screen background and the bright source documents.

Figure 157 illustrates an example of a very badly arranged workstation; unfortunately this picture cannot be considered an exception!

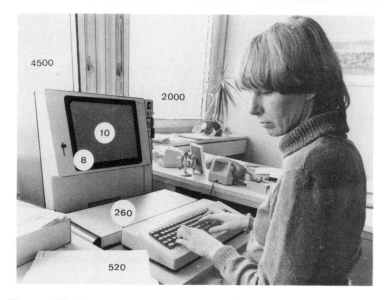

Figure 157. Excessive luminance contrasts in the visual environment of a VDT operator.
The figures in the circles indicate the measured luminances expressed in cd/m². Screen and source document have a contrast ratio of 1:50, screen and window of 1:450.

3:1 or 10:1?

According to the general guidelines the contrast ratio between the dark background of a screen and a well illuminated source document should not exceed 1:3. It is obvious that this recommendation cannot be realized at the majority of VDT workstations with bright characters and dark background. This situation induced several authors to re-examine the validity of the '1:3 rule'.

Haubner and Kokoschka [105] observed no decrease in performance up to luminance contrasts of 1:20 between screen and source document. Rupp [224] pointed out that the contrast between bright characters and source document might be more important than the contrast between screen background and source document.

Although not all problems of spatial and temporal differences of surface luminances in the visual environment of VDT operators have been solved, it is certainly realistic and reasonable to make the following propositions:

Recommended contrasts at VDT workstations

The luminance contrast between the dark screen (with bright characters) and the source document should not exceed the ratio of 1:10. All other surfaces in the visual environment should have luminances (reflectances) lying between those of the screen and the source document.

These recommendations are illustrated in Figure 158.

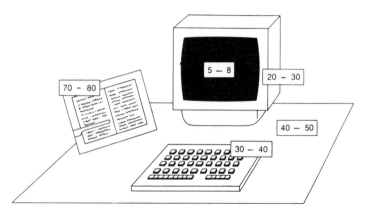

Figure 158. Recommended reflectances for a VDT workstation with a dark screen and a bright workstation environment.
The reflectance is the percentage of reflected light related to the luminous flux falling onto the surface concerned.

Contrasts with bright screens

Some terminals have dark characters on bright backgrounds ranging between 50 and 100 cd/m². It is obvious that such bright screens do not raise the problem of excessive contrast ratios with source documents or other bright surfaces in the visual environment, an undoubted merit!

Reflections on screen surfaces...

The surface of the screen is made of glass which reflects about 4% of the incident light; this is sufficient to reflect clear images of the office surroundings such as lights, the keyboard or the image of the operator.

One part is reflected from the glass surface of the screen; it

Figure 159. The reflected image of a window behind the back of the operator is superimposed on the screen text and disturbs reading.

produces a mirror-like reflection of the surroundings. The other part is reflected from the phosphor layer, producing a veiled and diffuse reflection of a light source. Figure 159 shows a common example of a reflected window on the screen.

...produce glare or annoyance

Bright reflections can be a source of glare and image reflections are annoying, to say the least, especially since they also interfere with focusing mechanisms; the eye is forced to focus alternately on text and reflected image. Thus reflections are also a source of distraction. Bright reflections on the screen are often the principal complaint of operators.

Positioning of VDT workstations

The most effective preventive measures are the adequate positioning of the screen with respect to lights, windows and other bright surfaces. (Other protective measures, such as adjustment of screen angle and anti-reflective devices on the screen surface are discussed in *Ergonomics in Computerized Offices* [85].)

If the light source is behind the back of the VDT operator it can easily be reflected on the screen and cause reflected glare. If it is in front it can cause direct glare. These conditions are illustrated in Figure 160.

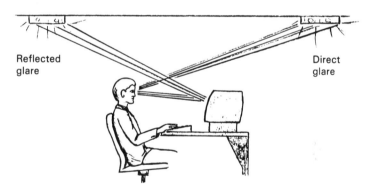

Figure 160. Light sources behind the operator create a risk of reflected glare; lights in front of the operator cause direct glare.

...with respect to light fixtures...

Light fixtures directly above the operator can dim the characters with blurred reflections generated in the phosphor layer. Thus it is preferable to install the light fixtures parallel to and on either side of the operator−screen axis.

...and with respect to windows

In offices, windows play a similar role to lights: a window in front of an operator disturbs through direct glare; when behind it produces reflected glare. *For this reason the VDT workstation must be placed at right-angles to the window.* In offices with only one or perhaps two parallel window walls this is an efficient protective measure. Figure 161 shows an ideal

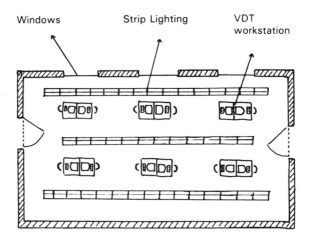

Figure 161. **VDT workstations should be arranged at right angles to the window.**
Plan of an office layout with one window wall.

arrangement of VDT workstations in relation to a single window wall.

Cover windows

In offices with two or more window walls some form of window covering is to be used. Windows must also be covered at night because the reflections of interior office lights may cause glare. Two types of window coverings can be useful:

Louvers or mini-blinds. Horizontal as well as vertical louvers can be used. Their purpose is to occlude the window on a bright day or to absorb light from indoor sources at night.
Curtains. They are efficient; preference should be given to material of low reflectance of about 50%.

Finally, there is the possibility of placing intermediate screens between the VDT workstation and bright windows. Such a screen should not have a higher reflectance than about 50%.

Appropriate light fixtures

The best light fixtures for offices with VDTs are not the same as those for traditional offices. Fixtures which provide a great deal of mainly horizontally directed light should be avoided since such light illuminates the vertical screen and generates reflections on it. It is advisable to use fixtures which provide a confined primarily downward distribution of light, with either built-in louvers, curved mirrors or prismatic pattern shields. The luminous flux angle should not exceed 45° to a vertical. Suitable and well arranged light fixtures are shown in Figure 162. Such fixtures cause neither direct nor reflected glare since screen and keyboard are in a shadow area.

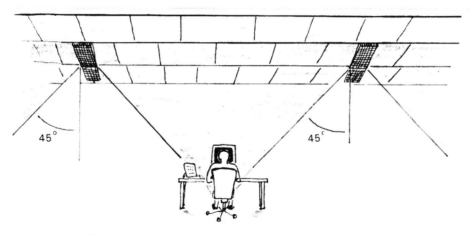

Figure 162. Ceiling lighting with a prismatic pattern shield generating a cone of light with an angle of 45° to the vertical.

Indirect lighting

Some lighting engineers suggest suspending the luminaires from the ceiling, thus permitting the lowered fixtures to direct the greater part of the light upwards to the ceiling. Standard lamps emitting the full light upwards towards the ceiling and the upper part of the walls are also used in some offices. These lighting systems may produce a pleasant aesthetic effect but have the drawback of bright ceilings and walls, which, in turn, may cause bright reflections on the screen.

Task lighting

Given a low general illumination level supplementary lighting may be provided for the source documents, especially when they have a low readability. It is important that such task lighting be confined to the area of the source documents. In order to avoid reflections on neighbouring VDT workstations the task lamps should not be transparent at the sides.

19. *Noise and vibration*

19.1. Perception of sound

Sound perception

The physiological processes of perception of sound are essentially the same as those already discussed for visual perception. In this case the inner ear provides the 'interface' at which soundwaves are converted into nerve impulses along the auditory nerve. *The actual perception of sound is the integration and interpretation of these sensory impulses in the brain, or more precisely in the auditory cortex.*

Perception of sound is not a faithful reproduction of the whole band of frequencies, which is simply 'played' in the brain. This fact is especially important in people's reaction to noise, which varies greatly from person to person. What is noise to one may be music to another. Another example of varying perception is the assessment of loudness in relation to pitch. In practice low-pitched sounds seem less loud than shrill ones, even though the energy content may be the same.

Sounds

Any sudden mechanical movement sets up fluctuations in the air pressure which spread out as waves, just like waves when water is stirred. As long as these variations of pressure occur with a regular frequency and intensity, the human ear reacts to them as sounds. *The extent of pressure variation is the sound pressure, and this determines the intensity of the sensation.*

The frequency of a sound is the number of fluctuations or vibrations per second, expressed in herz (Hz), subjectively perceived as pitch. Most noises contain a mixture of sounds of different frequencies; if high frequencies predominate, we regard it as a high-pitched noise, and vice versa.

The decibel

The physical unit of sound pressure is the micropascal (μPa). The weakest sound that a healthy human ear can detect is about 20 μPa. This pressure wave of 20 μPa is so low that it causes the membrane in the inner ear to deflect by less than the diameter of a single atom! But the ear can also tolerate sound pressures up to more than 1 million times higher. The range of hearing encompasses every sound from the gentle murmur of a stream to the scream of a jet engine.

To accommodate such a wide range in a practical scale, a logarithmic unit, the decibel (dB) was introduced.

The decibel scale uses the hearing threshold of 20 μPa as a reference pressure. Each time the sound pressure in μPa is multiplied by 10, 20 dB are added to the decibel level, so that 200 μPa corresponds to 20 dB. 1 dB is the smallest change the ear can distinguish; a 6 dB increase is a doubling of the sound pressure level, although a 10 dB increase is required to make it sound twice as loud.

Sound pressures are recorded logarithmically by using the sound level L (sound intensity level), according to the formula:

$$L_{dB} = 20 \, \log\frac{p_x}{p_0}$$

where L_{dB} = sound level in dB; p_x = sound pressure in μPa; and p_0 = basic sound pressure, internationally fixed at 20 μPa.

Pitch and loudness

As we have seen, the apparent loudness of a sound depends a good deal on its pitch or frequency. The human ear is sensitive to sounds in the frequency range from 16 to 20 000 Hz, a span of nearly nine octaves. Sounds pitched below 16 Hz (infrasonic) are perceived as vibrations: above 20 000 Hz they are ultrasonic and are used therapeutically in medicine. *Low-pitched sounds seem much less loud than high-pitched ones.* This is shown very clearly by the curve of auditory threshold for different frequencies, the lowest curve in Figure 163. This curve shows that the greatest sensitivity lies in the range 2000−5000 Hz, although the pitch of the human voice is much lower than this, mainly between 300 and 700 Hz.†

Curves of equal loudness

As long ago as 1933 Fletcher and Munson [71] plotted curves of equal loudness in relation to sound pressure and frequency. For this purpose they worked from an arbitrary base of 1000 Hz and tones of higher and lower frequencies, determining the sound pressures necessary to give the research subjects the impression of equal loudness. In this way they obtained curves of equal subjective loudness which were afterwards converted to a scale of phons. Robinson and Dadson [214] carried these studies further, and their results have been incorporated in the ISO Standard since 1957. They are displayed in Figure 163.

These curves of equal loudness, on the scale of phon values, are valid only for pure tones; they no longer agree with subjective impressions of loudness if the noise includes many different frequencies. Since nearly all noises, and many signal sounds, are a mixture of frequencies, the use of phon values as

---

† Most vowels are pitched below 1000 Hz but consonants may be pitched up to 10 000 Hz, especially if they are sibilant.

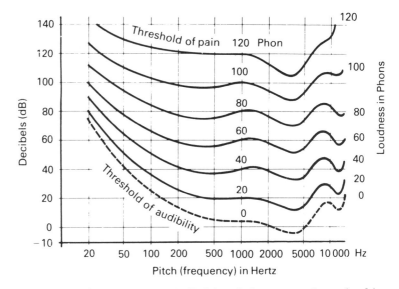

Figure 163. Sound level (in decibels) and the curves of equal subjective loudness (in phons).
The lowest curve indicates the threshold of audibility, the least sound level that can be perceived. After ISO [128].

a measure of loudness has become obsolete. Nevertheless these curves of equal loudness still retain their value for assessing the effects of different frequency ranges on the human ear.

Weighted noise level

Nowadays the so-called weighted sound levels have come into use as measures of loudness. Weighted sound level is essen-

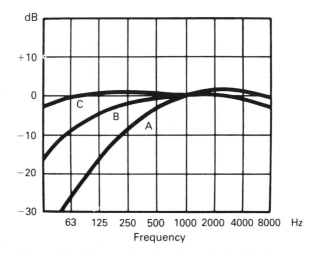
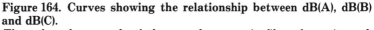

Figure 164. Curves showing the relationship between dB(A), dB(B) and dB(C).
They show how much of the sound energy is filtered out in each frequency range.

tially a process of filtering out the sound energy in the lowest and highest frequencies, at which, according to the curves of equal loudness, sensitivity is less. Hence sound pressure has less significance in these frequency ranges. The term 'weighted noise level' has a similar derivation. Figure 164 sets out three weighted curves in dB, (A), (B) and (C), that are in current use.

The curve of weighted sound level in dB(A) is most often used, because many psychological studies have shown that *noise levels measured in dB(A) give a reliable assessment of subjective disturbance from noise.*

Other units of measurement which are useful in particular circumstances, are discussed later.

Hearing organs   A sensation of hearing is produced when sound waves pass through the external auditory passages, into the inner ear, where the energy from sound pressure is converted into nervous impulses. These travel along the auditory nerve to the brain, where the sound is 'heard'.

The most important components of the hearing organs are shown in Figure 165. Sound waves cause the eardrum or tympanic membrane to vibrate, and these vibrations are transmitted by the auditory ossicles to the inner ear. Here fluids (the perilymph and enolymph) transmit the vibrations along the cochlea and back to the round window. The so-called basilar membrane divides the cochlea longitudinally into two chambers, and contains the organ of Corti with its sound-sensitive cells; it is these which convert the pressure waves into nervous impulses. Each cell is sensitive to a particular range of frequencies, that is to sounds of a certain range of

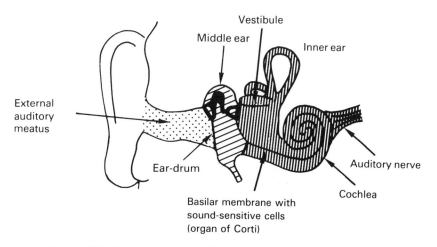

Figure 165. The anatomy of the ear.
The shaded parts belong to the inner ear, in which the cochlea detects the sound, whereas the curved tubes overlying it (the vestibule) perceive acceleration and balance.

pitch, and passes its stimulus to a single nerve fibre, which conveys it to the brain.

The basilar membrane has a length of about 30 mm. The sense cells at the apex of the cochlea are stimulated by low-frequency sounds, whereas those at the large entrance (adjoining the basal plate of the stapes) respond to the high frequencies. At this point, movements of the stapes (the last of the auditory ossicles) create a series of waves, and the distance from the stapes to the crest of each wave (i.e., wavelength) is a function of its frequency. High notes set up short waves, with their crests near to the beginning of the cochlea; low notes create longer waves, with their crests nearer the apex of the cochlea. Hence the point at which the basilar membrane receives maximum pressure depends on the frequency (i.e., the pitch) of the sound.

Müller-Limmroth [192] compares this phenomenon with the breaking of waves in surf; short waves (high frequencies) 'break' early, at the beginning of the cochlea, whereas longer waves (low frequencies) travel progressively further before 'breaking'. In this way the inner ear carries out a sort of 'frequency analysis' of the incoming sound waves, and transmits information to the brain of all the frequency components making up the sound. The cerebral cortex integrates these components again so that we are conscious only of the sound as a whole, and not of its constituent parts.

Auditory
pathways

The nervous impulses generated by the sound waves travel along the auditory nerve to the brain, entering into the brain

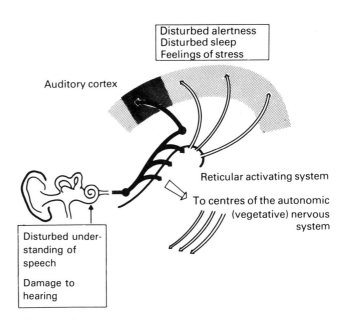

Figure 166. The auditory pathways (solid arrows) and the accessory tracks (open arrows) which set up secondary effects of noise.

stem or medulla, and passing through two synapses or nerve connections to the auditory sphere of the cerebral cortex. Here the separate nerve impulses are localized, as if the cochlea of the ear were spread out over the cerebral cortex. This is where the brain finally integrates these impulses into an impression of the sound. Figure 166 shows these auditory pathways diagrammatically.

It should be emphasized once again that the conscious hearing is a phenomenon of the brain, more precisely of the cerebral cortex. The inner ear and the auditory pathways are no more than the transmitting mechanism, the 'interface' between the atmospheric sound and the conscious perception of the brain.

Side-effects

Figure 166 also shows that between the two synapses, mentioned above, nerve fibres lead out to the reticular activating system. From here fibres run to the entire conscious sphere of the cerebral cortex, so that excitation of the system by incoming acoustic signals may induce alarm throughout the consciousness of the individual, disturbing sleep, reducing concentration when awake, and producing other distressing symptoms. This kind of 'alarm call' has a biologically important role in alerting a person, giving him or her an opportunity to interpret the noise and react suitably to it. Here is an example: a pedestrian is strolling dreamily along a beautiful country road, and the noise of an approaching motor car grows steadily louder. At first the pedestrian remains unconscious of this, but when the noise has reached a certain level the reticular activating system is excited and sends a signal to the brain. The pedestrian is immediately alerted, so that he can:

1. Take conscious notice of the sound.
2. Interpret it.
3. Get out of the way of the car.

Hearing is an alarm system

It is clear that hearing has two principle functions:

1. To convey specific information, as *a basis for communication between individuals*: this function is highly developed in Man.
2. As an alarm system: by activating secondary pathways leading to the brain *it plays an essential part in waking, increased alertness, and finally alarm.*

The alarm function of the sense of hearing may be used to advantage in planning transport and in industry. It is essential to recognize dangerous situations quickly, whether supervising a control panel or a switchboard, driving a locomotive, or flying an aeroplane. To this end a suitable combination of acoustic signals and visual aids is needed. The acoustic signals serve to 'alert' the brain, and the visual aids convey the necessary information.

19.2. Noise load

Definition

The simplest definition is that *noise is any disturbing sound.* In practice we call it simply 'sound' when we find it not unpleasant, and 'noise' when it annoys us. Hawel [106] defined it more precisely as follows: "Sound only becomes burdensome if the person affected feels that it is discordant, i.e. if it does not harmonise with his intentions at that particular moment". This definition is particularly apt when applied to noise at work.

Measurements of noise load

The *noise load* is the extent of the noise, as measured in physical units, taking account of all the acoustical factors over a given time. Various pieces of research work have shown that the *noise level* is not the only factor operating, but that the frequency with which the noise occurs and other quantities contribute to the total noise load.

These studies have led to the development of units of measurement which combine various components of the noise load into one quantity, so that it is then possible to characterize the noise load at a particular point by a single figure.

Two such units are important in assessing noise problems at the workplace:

1. *The equivalent level of sustained noise (continuous sound level);*
2. *The summated frequency level.*

The equivalent noise level L_{eq}

The equivalent level of sustained noise (L_{eq}) expresses the average level of sound energy during a given period of time (i.e. the energy level). This quantity is an integration of all the sound levels which vary during this time, and so compares the disturbing effect of the fluctuating noises with a continuous noise of steady intensity.

The summated frequency level

The summated frequency level is measured with a sound level indicator and a frequency counter, operating over a given time. Commonly used units of sound measurement include:

1. L_{50} *(average noise level).* '$L_{50} = 60$ dB' means that the level of 60 dB was reached or exceeded during 50% of the relevant time.
2. L_1 *(peak noise level).* '$L_1 = 70$ dB' means that the level of 70 dB was reached or exceeded for 1% of the time.

These two levels L_{50} and L_1, are related to the equivalent noise level by the following approximation:

$$L_{eq} = L_{50} + 0 \cdot 43 \, (L_1 - L_{50}) \simeq \frac{L_{50} + L_1}{2}$$

Sources of noise	Disturbing noise may be either *external*, coming from outside the building, or *internal*, generated within the building itself. The most important sources of external noise are traffic, industry, building and neighbours.

The most important sources of internal noise in factories are machines, engines, compressed air, milling machines, stamping machines, looms, sawmills and any other pieces of noisy machinery. In addition to this, the office has its own internal noise which comes from telephones, typewriters and keyboards, calculating machines, printers and people walking about and talking.

Street noise

External noise is a disturbing factor in offices, drawing offices, conference rooms and schools. Table 34 lists some of the noise levels to be expected in premises alongside a road:

Table 34. Noise levels from street traffic, given as L_{eq} in dB(A).
Measuring point in front of the window: if this is opened, the noise level in the room will be 5—10 dB(A) less than that recorded outside.

Traffic density	L_{eq} dB(A)	
	By day	By night
Heavy (main road with through-traffic)	65—75	55—65
Moderate	60—65	50—55
Light (local street)	50—55	40—45

Industrial noise

Internal noise in factories is always a very variable factor. It may be continuous or intermittent, and may take the form of banging, clattering, rattling or whistling. Table 35 shows some of the peak levels that may be attained.

To assess the extent of the danger to hearing from such noise we need to consider the average level of L_{eq} during the

Table 35. Peak noise levels, in dB(A).

Source	Noise level, dB(A)
Rifle-shot; engine test bench	130
Pneumatic bore-hammer	120
Pneumatic chisel	115—120
Rocking sieve; chain-saw; compressed air rivetter; electric cutter; compressed air hammer	105—115
Milling or weaving machine; crosscut saw; stamping machine; boiler-house; weaving shed	100—105
Electric motor; rotary press; wire-drawer; sawmill; composing room; bottle-filling machine	90—95
Toolmaking machine (running light)	80

Table 36. Equivalent noise level L_{eq} in dB(A) at various work places, *calculated over an 8-hour shift*

Department or machine	L_{eq} dB(A)
In a flexible-tube factory	
At yarn-spinning machine	95
In the spinning room	90
At the weaving machine	95
In the weaving shed	95
In a soft-drinks plant	
At the mixing plant	95
At the washing check-point	100
At the automatic sealing machine	100
At a can-filling machine	90

8-hour shift. According to ISO TC43 [128], the measurement should be in dB(A), and equivalent level of sustained noise (L_{eq}) calculated for intermittent noise.

Table 36 gives a few examples of L_{eq} for an 8-hour day.

Noise in offices

Concentrated mental work, or jobs at which understanding of speech is important are 'noise-sensitive' occupations, and even if the noise level is comparatively low it can be disturbing.

Noise levels to be expected in offices are listed in Table 37. The peak noise level L_1 lies in the zone between 56 and 65 dB(A), which is about 4−8 dB(A) above the average noise level L_{50}.

Nemecek and Grandjean [198] studied noise problems in 15 large offices in 1970, at a time before VDTs had been introduced. The results showed equivalent noise levels, L_{eq}, of 52−62 dB(A) with peak noise levels, L_1, of 56−65 dB(A). It was concluded that large open offices are rather noisy.

Table 37. Noise levels usual in offices.

Type of office	Average L_{eq}, in dB(A)
Very quiet, small offices and drawing offices	40−45
Large, quiet offices	46−52
Large, noisy offices	53−60

Noise emission of office machines

The advent of VDTs brought a number of additional machines into the office: printers, plotters, calculators, and so on.

In Table 38 the noise emissions of some modern office machines are presented.

Table 38. Approximate noise emissions of office machines.
Measurement position about at the head level of operators.

Machine	Noise emission, in dB(A)
Matrix and daisy wheel printers:	
Basic noise	73−75
Peak levels	80−82
Matrix printer with hood:	
Peak levels	61−62
Inkjet printer:	
Basic noise	57−59
Peak levels	60−62
Laser printer	No measurable noise
Cooling fan of VDT	30−60
Old typewriter	c. 70
Modern electronic typewriters	c. 60
Two electronic typewriters face to face	68−73
Copying machines	55−70

Matrix and daisy wheel printers are a torture!

It is clear that matrix and daisy wheel printers produce by far the highest noise emissions. Their peak levels, which are highly repetitive, interfere strongly with speech communication and are very annoying, particularly to those who do not benefit from their use. There are some possibilities of reducing this noise: hoods deaden the noise emission by about 20 dB(A); unfortunately, many employees dislike having to open and shut the hood each time, except when a great amount has to be printed for a long period of time. Where several printers are being used, one should consider locating them in a separate room. The panels which are used to wall off units in large open offices reduce the noise emissions only by a few decibels; they do not suffice to damp noise sources exceeding 70 dB.

Inkjet and laser printers are noiseless machines and are greatly appreciated except by those who have to pay the bill!

The VDT itself is noiseless, but not the cooling fans

The VDT itself is nearly silent but many makes have built-in electric fans to cool the unit from the heat produced by the CRT†. As long as the noise of the electric fans is below 40 dB(A), it will hardly be noticed, but when it exceeds 50 dB(A), it will become extremely irritating. Personal computers, located in a quiet closed space, frequently create this noise problem.

Modern electronic typewriters as well as the keyboards attached to VDTs do not produce excessive noise and are seldom a source of complaints. Most modern copying machines, too, have rather low noise emissions and give no cause for complaint either.

---

† Cathode Ray Tube.

19.3. Damage to hearing through noise

Hearing losses

Strong and repeated stimulation from noise can lead to loss of hearing, which is only temporary at first, but after being 'deafened' repeatedly some permanent damage may occur. This is called *noise deafness*, and is brought about by slow but progressive degeneration of the sound-sensitive cells of the inner ear. The louder the noise, and the more often it is repeated, the greater the damage to hearing. Moreover, it is well known that noise consisting of predominantly high frequencies is more harmful than low-frequency noise. Intermittent noise, such as hammering, is more harmful than continuous noise, and a single very loud noise — e.g., a detonation or an explosion — can damage the ears immediately.

Individual sensitivity to noise varies greatly from one person to another. Some who are particularly sensitive may suffer permanent deafness after only a few months, whereas less sensitive people may not show the first symptoms until after many years' exposure. Noise deafness starts with the higher frequencies, at about 4000 Hz, and extends only gradually to the lower frequencies. At first the worker is unaware of it, and only gradually notices his loss of hearing when it begins to involve the lower frequencies. Noise deafness is progressive, and often combines with the deafness of increasing age, or is mistaken for the early onset of the latter. In most industrial countries, noise deafness is one of the occupational hazards of working life.

Audiometry

Nowadays the extent of loss of hearing is usually measured by means of so-called pure tone audiometry, which determines the threshold of hearing of pure tones of various frequencies. The result is recorded on an audiogram, which shows in dB how much the threshold of hearing has been raised for each frequency. Such an audiogram of impairment of hearing by noise is shown in Figure 167.

Temporary loss of hearing

In this context we must return to temporary noise deafness, already mentioned above. This phenomenon is characterized by the *hearing returning to normal;* known as the *'temporary threshold shift'*. Reversible deafness has been the subject of intensive study by Kryter [155], Ward [266] and others. These authors have found that there is a very close relationship between temporary and permanent deafness, and that the results for temporary loss of hearing allow many conclusions to be drawn that are of more general application. The most important of these may be listed as follows:

1. Noise up to 80 or 90 dB causes only slight shifts in the threshold of hearing, 8 or 10 dB only, but if the noise is increased to 100 dB, the threshold goes up by 50 to 60 dB.

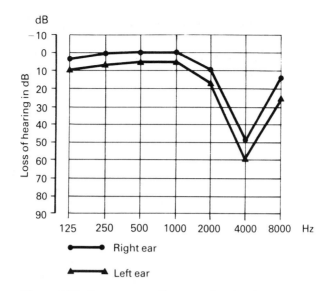

Figure 167. Pure-tone audiogram showing impairment of hearing by noise.
The zero line is the normal threshold of hearing. The loss of 50—60 dB at 4000 Hz is characteristic of hearing damage by noise.

2. The temporary shift in the threshold of audibility is proportional to the duration of the noise. For example, a 100 dB noise for 10 minutes produces a shift of 16 dB, and after 100 minutes, one of 32 dB.
3. The time taken for hearing to return to normal is also proportional to the intensity and duration of the preceding noise. The time of restitution is about 10% longer than the duration of the noise.
4. Alternate noisy and quiet periods produce less temporary deafness.

Age deafness   The threshold of hearing rises progressively with age, and loss of hearing is greatest in the higher ranges of frequency and more pronounced in men than in women. Taking a frequency of 3000 Hz as standard, the loss of hearing to be expected at various ages is as follows:

50 years: 10 dB
60 years: 25 dB
70 years: 35 dB

The audiogram of age deafness differs from that of noise deafness in that the loss of hearing increases progressively as the frequency is raised, so that the highest frequency still audible shows the greatest shift in its threshold. The characteristic dip in the curve at 4000 Hz is not evident in cases of age deafness.

Older workers often show the combined effects of age deafness and noise deafness, and it is very difficult to distinguish the two.

*The risk of loss
of hearing*

From the evidence of many comparisons between exposure to noise and the frequency of impaired hearing, it is now possible to estimate the risk of hearing damage in noisy factories. The ISO publication [128] sets out in a comprehensive table this risk in relation to age, duration of exposure, and the intensity of the noise (expressed as L_{eq} for a 40-hour week). A simplified extract from this report is given in Table 39.

These figures show that the risk of damage increases both with sound intensity and duration of exposure, *the damaging intensities being those above approximately 90 dB(A)*.

Factory workers are often exposed to noise that varies widely, and it has been shown that interruptions, or periods of relative quiet, reduce the risk of damage to hearing.

To assess the extent of such risk, the equivalent noise level over the 8-hour working day must be calculated. The relation between length of exposure and intensity of sound to create the same degree of risk is as follows:

Hours	dB(A)
8	90
6	92
3	97
1.5	102
0.5	110

Table 39. **The risk to hearing, as a presumptive percentage of the work force.**
The percentages will be increased by several units with age.

L_{eq} in dB(A)	Length of exposure in years		
	5 (%)	10 (%)	20 (%)
80	0	0	0
90	4	10	16
100	12	29	42
110	26	55	78

As far as current knowledge of noise damage goes, *the following may be proposed as a limiting value*: L_{eq} *for an 8-hour day not to exceed 85 dB(A)*. Most of the limits in use today approximate to this, although they are often more precisely defined and formulated [128, 262].

19.4. Physiological and psychological effects of noise

Figure 166 shows the auditory tracts and how they are linked with the activating and alarm-sensitive structures of the brain. Here lies the explanation of many of the effects of noise and the spread of a state of activation or alarm throughout the consciousness. This may result in:

1. *Impaired alertness*
2. *Disturbance of sleep*
3. *Annoyance.*

At the same time this activation affects the autonomic centres and produces the so-called *vegetative effects* in the internal organs. Finally, a special problem of noise is that it makes it *difficult to understand what people say*. The ergonomic aspect of these various problems will now be discussed.

Understanding of speech
We all know from our own experience that the sensitivity of the ear to one particular sound — say the voice of a colleague — becomes less and less as the ambient noise increases. The ability to pick out one particular sound from the rest depends on its auditory threshold, which rises linearly with the sound intensity up to 80 dB. Where human speech is concerned, however, it is not enough to hear the pure tones. The message must also be understood, and to do this requires a very special discriminating ability in the ear. A critical factor is correct hearing of the consonants, which are much softer sounds than the vowels.

Since the understanding of words and sentences depends greatly on individual intelligence, as well as on familiarity with the test language, research workers into the damaging effects of noise have used *syllables* as the criteria. A speaker utters a succession of meaningless syllables and records what proportion of the total number are correctly understood. This has become the standard method for research in this field.

It has been shown that a considerable understanding of sentences and their meaning is possible without understanding all the separate syllables. Figure 168 shows the ratio between syllable comprehension (S) and sentence comprehension (W).

It can be seen from this graph that with a syllable comprehension of only 20% it is still possible to understand nearly 80% of the sentences; if half of the syllables are understood ($S=0\cdot5$), then about 95% of the sentences are intelligible.

Speech comprehension in a workroom depends very largely on the loudness of the voice concerned and the level of background noise. Figure 169 shows the relationship between syllable comprehension (S), noise level (N) and voice level (P) (solid lines).

In addition, the dotted lines join together points at which there are equal differences between sound pressure of noise and sound pressure of speech: $P-N=20, 10, 0$ and -10 dB respectively. It is apparent from this graph that a syllable

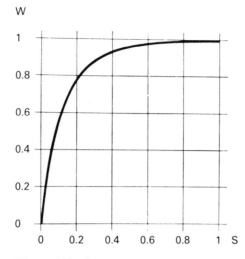

Figure 168. Comprehension of syllables (S) and complete sentences, or sense (W).
Each is plotted as a decimal fraction of the total number of syllables or sentences offered for comprehension.

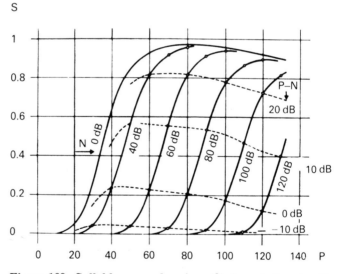

Figure 169. Syllable comprehension (S) in relation to the sound-pressure (P) of speech and the noise level (N) in the room.
The dotted lines connect all those points at which P – N is 20, 10, 0 or –10 dB.

comprehension of 40–56% is possible as long as $P - N = 10$ dB. According to Figure 168, this implies 93–97% comprehension of sentences, or sense (*W*). *Experience shows that this level of comprehension is good enough for most factories and offices, and therefore speech comprehension is considered to be unimpaired, as long as the background noise level is at least 10 dB below the level of the speaking voice.*

If, however, we are concerned with an exchange of verbal information, on a subject that is unfamiliar, with difficult new words, then a higher level of syllable comprehension is necessary. It has been demonstrated that in these circumstances syllable comprehension must be as high as 80%, and this requires a difference of 20 dB between the pressure levels of voice and background noise (Figure 169).

Sound pressure of noise

The normal speaking voice, indoors, at a distance of 1 metre, operates at the following pressure levels in dB:

Quiet conversation:	60−65
Dictation:	65−70
Speaker at a conference	65−75
Delivery of a lecture:	70−80
Loud shouting:	80−85

If the voice has to be used frequently to dictate, or to convey information in the course of a person's employment, it should not exceed 65−70 dB at a distance of 1 m. If this is to be understood clearly and without strain, the background noise level must not exceed 55−60 dB, and if the verbal communication is more difficult to understand, e.g., contains many strange words, or unfamiliar names, then the background noise must not exceed 45−50 dB.

If offices or workrooms with these requirements are situated immediately on roads with moderate or heavy traffic densities, then a maximum permitted background noise of 55−60 dB cannot usually be adhered to, especially in summer when windows are open. In these circumstances noise levels of 70−75 dB are often reached in offices. Offices in summer need more effective protection against traffic noise, in the form of air-conditioning, which allows the windows to be kept permanently closed. In towns, traffic noise is the most important reason for installing air-conditioning.

Effects on performance

Exposure to noise has little effect on manual work, whereas we all know from experience that thought and reflection are more fatiguing in a noisy environment than in a quiet one. Sports coaches in particular, know that disciplines and all movements that call for intense nervous concentration are adversely affected by noise. Many everyday examples show that noise impairs concentration, and one is led to assume that under such conditions performance and output will also suffer. It is interesting, though, that what seems to be a truism in everyday life is only partly confirmed by either experiments or field studies. Research into the effects of noise on either mental or psychomotor performance has given very contradictory results: noise is just as likely to improve performance as to make it worse. We must conclude that *a decline in performance can be attributed to the effects of noise, but with important reservations.*

*Improving
performance*

Noise can even be positively stimulating in the right circumstances. Performance may be *improved* if the work is boring, and perhaps also in situations where there are many other distractions. Sometimes, if these distractions can be concentrated on one dominant noise, even mental performance may be improved.

*Loss of
performance*

However, these examples of improved performance must be set against more pieces of research that have shown performance to be adversely affected by noise. For example, Broadbent [28, 30] and Jerison [132] have observed a decline in performance during exacting tests of sustained vigilance.

To summarize, we can say of the deleterious effects of noise in performance that:

1. *Background noise often interferes with complex mental activities, as well as certain kinds of performance that make heavy demands on skill and on the interpretation of information.*
2. *Noise can make it more difficult to learn certain kinds of dexterity.*
3. *Many studies have shown that high levels of noise (over 90 dB as a rule), either discontinuous or unexpected, can impair mental performance.*

Field studies

The effects of noise have occasionally been studied under industrial conditions. Thus Wisner [275] reports the following experiences:

1. In one machine shop a reduction of about 25 dB in the noise level led to 50% fewer rejected pieces.
2. In an assembly shop, a reduction of the noise level by 20 dB raised production by about 30%.
3. In a typing pool, a reduction of noise level by about 25 dB was accompanied by a 30% reduction in typing errors.

Such results must always be interpreted with caution, however. Wisner himself hints that the new situation in the factory may have a psychological effect on the workers.

*Effects of noise
in offices*

An interesting example of this was revealed during the previously mentioned study in 15 open-plan offices. The summated frequency level was measured in each of the offices, and at the same time 519 office workers were questioned about their experiences and opinions. About 8000 measurements of noise levels gave average values (L_{50}) between 38 and 57 dB(A), and peak noise levels (L_1) between 49 and 65 dB(A). The distribution of replies to the question, "What kinds of noise disturb you?" is shown graphically in Figure 170.

It is clear that *'talking' or 'conversation' is the disturbance most often complained of.* Many of those questioned added

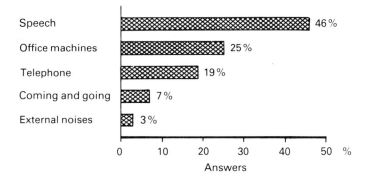

Figure 170. **Frequency distribution of replies to a question about the sources of noise.**
411 people questioned, some mentioning more than one source, giving a total of 762 replies = 100%. After Nemecek and Grandjean [198].

that it was not the loudness of the conversation that disturbed them, but its content!

These indications were confirmed by the results of correlations calculated between noise levels and the incidence of rating noise as 'very disturbing'. Indeed, no correlation could be found between noise levels and disturbance. This means that the disturbance due to office noise is almost independent of the measured noise levels.

Information content of talks is distracting

These results led to the conclusion that the *conversations of other people are distracting not so much through their sheer loudness as through their information content.*

Conversations of other people, which are usually not very loud, can be masked by the general background noise caused by office staff, rustling paper, walking, typing, office machines and the room ventilation. One may conclude that a certain average noise level, which covers up the conversations of others, would be appreciated by many office employees. This is why in some large open offices fixtures have been installed, producing a regular, constant background noise, called 'sound conditioning', intended to mask conversations. Since opinions about sound conditioning have been rather controversial, it is not yet possible to draw definite conclusions.

Physiological stress from noise

Many physiological studies have shown that exposure to noise produces:

1. Raising of the blood pressure.
2. Acceleration of heart rate.
3. Contraction of the blood vessels of the skin.
4. Increase in metabolism.
5. Slowing down of the digestive organs.
6. Increased muscular tension.

All these reactions are symptomatic of a spreading state of alarm, which is generated and controlled by a state of increased stimulation of the autonomic nervous system. This is actually a defensive mechanism which prepares the whole body for facing possible danger, by being ready for fight, flight or defence. It should not be forgotten that throughout the animal kingdom the sense of hearing is primarily an alarm system, and this basic function still remains even in the human organism.

Waking effects

It is essential for the maintenance of good health that the stresses of the day should alternate with the restorative powers of sleep.

The acoustic sense is the most effective awakener from sleep. When the eyelids are closed, optical stimuli are largely excluded, whereas the sense of hearing is only slightly muffled during sleep. It still retains its primary function as an alarm system.

Experience shows, in fact, that familiar noises are less likely to awaken a sleeper than unfamiliar ones. People who live close to a railway are not awakened by passing trains, whereas an anxious mother awakes if her child coughs or breathes strangely. Obviously the human brain can 'tune in' to certain sounds and react to those by awakening, while ignoring others, but has little defence against totally unpredictable sounds, when neither their nature nor their timing can be foreseen. Such noises have a powerful waking effect.

Noise may either awaken people completely, or rouse them into a twilight sleep. This is particularly likely if the noise is repeated. Since twilight sleep is not as restful as deep sleep, it is undesirable, and so intermittent noises may be said to impair the quality of sleep even if the subject does not wake up completely.

Noise and sleep

The effects of noise on the duration and quality of sleep under experimental conditions has been studied in detail with the aid of the electroencephalograph. Studies up to now have shown that a noisy environment:

1. Seriously curtails the total time asleep.
2. Cuts down the amount of deep sleep.
3. Increases the time spent awake, or in light sleep.
4. Increases the number of waking reactions.
5. Prolongs the time to fall asleep.

Annoyance

Everyday experience teaches us that many noises have effects of an emotive nature on people. These arouse feelings and sensations, and so are strongly subjective. They are reckoned among the psychological effects of noise. Not all noises or sounds are burdensome. Natural sounds, such as the rustling of leaves or the murmur of a stream, are pleasant. Noise may also be agreeable in itself, if it drowns other sounds that are not pleasant. On the other hand there are many kinds of noise, or noisy situations, which people regard as being subjectively unpleasant and burdensome. The nature and extent of the burden depends on a number of subjective and objective factors, the most important of which are as follows.

1. The louder the noise, and the more high frequencies it contains, the more people are affected by it.
2. Unfamiliar and intermittent noises are more troublesome than familiar or continuous sounds.
3. A decisive factor is a person's previous experience of the particular noise involved. A noise which often disturbs one's sleep, which excites anxiety, or which interferes with what one is doing, is particularly burdensome.
4. A person's attitude to the source of the noise is often specially important. A motor-cyclist, a workman, a child or a musician is not disturbed by the noise generated by his own activities, whereas a bystander or someone not partici-pating is disturbed to an extent which depends on how much he dislikes either the sounds being produced, or the person producing them.
5. The extent of the disturbance by noise often depends upon what the person affected is doing, and what time of day it is. A working housewife is less disturbed by traffic noise and noise from the neighbours, all day long, than is her husband during his short midday break, when he wants to rest and relax. The rustling of papers is disturbing during a lecture, whereas out in the street it would pass unnoticed.

The burdensome feelings generated by noise are among its most important effects, because they are so widespread, and they must be regarded as the decisive factor in developing techniques for combating noise, and formulating regulations against it.

Becoming accustomed to noise

It is still not clear how far people can become accustomed to noise. Experience shows that there may be some degree of adaptation in certain circumstances, yet in others there is either no adaptation at all, or even the converse, an increasing sensitivity to noise. These phenomena all depend on so many external circumstances, and so many psychological factors, that it is still not possible to generalize. We can only infer from the continued increase in noise and its burdens that no adaptation will be possible. On the contrary, the limit of adaptability will be overreached more and more often.

Noise and health

The recuperative processes that are essential to health take place during night sleep, and during pauses of all kinds, interruptions of work and a person's leisure time.

If the irritating effects of noise on the autonomic nervous system are not confined to working hours, but extend into the hours of sleep, then this will upset the balance between stress and recuperation. Noise will then become a causative factor in states of chronic fatigue, with all its ill-effects on well-being, efficiency and the incidence of ailments.

According to the definition of the World Health Organization (WHO) health is a state of physical and mental well-being. If we take this definition as a basis for discussion, then not only must deafness, but also the frequent disturbances to sleep, delayed recuperation, and the daily repetition of the burdens of noise be reckoned among the hazards to health.

19.5. Protection against noise

Guidelines for office noise

Based on general experience, as well as on studies, the guidelines reported in Table 40 can be proposed for office noise. It is recommended that the noise in offices with more than 5−10 occupants should not be much lower or higher than the given ranges. For offices with one or two people the recommended values can be considered to be the upper desirable limit but lower noise levels would certainly be more favourable. Some experience indicates that the desirable background noise level is achieved in large offices of at least 1000 m^2, containing more than 80 people.

The recommended range of equivalent noise levels L_{eq} of 54−59 dB(A) will to some extent mask the conversations and telephone calls of others, while speech communication between two employees will remain undisturbed. This was recently confirmed by the results of a survey by Cakir *et al.* [37], who found the lowest prevalence of noise disturbances in computerized large open offices with a background noise between 48 and 55 dB(A). Below and above this range the incidence of complaints was clearly higher.

Table 40. **Guidelines for noise levels in large open offices.**
The office noise should principally be neither lower nor higher than the given ranges.

Noise measurement	Desirable range, in dB(A)
Equivalent noise level, L_{eq}	54−59
Mean noise level, L_{50}	
(approx. background noise)	50−55
Peak noise levels, L_1	60−65

Noise in factories

Noise in factories can be countered in the following ways:

1. *Protective planning.*
2. *Reduction of noise at source.*
3. *Insulation against reflection and scattering.*
4. *Personal sound protection*

Planning

The most important technological step in the battle against disturbing noise lies in the choice of building materials and in planning the subdivisions of the building. *Hence noise protection begins on the architect's drawing board.*

It is fair to assume that the noise level will decrease with increasing distance from its source, so it is advantageous that offices, drawing offices and any other places where mental work is carried on should be sited as far away as possible from the noise of traffic. If the factory itself is noisy, the noisy sections, too, should be as far away as possible from places where work is carried on that calls for concentration and skill; intervening rooms, used for packing and storing, will act as 'buffers'.

When considering the division between two rooms, account must be taken of the damping effect of walls, doors, windows and hatches; and Table 41 gives some examples of these.

Table 41. Sound-deadening effect of various building items.

Item	Damping effect, in dB(A)	Remarks
Normal single doors	21−29	Speech clearly understandable
Normal double doors	30−39	Loud speech still understandable
Heavy special doors	40−46	Loud speech still audible
Window, single glazing	20−24	
Window, double glazing	24−28	
Double glazing with felt packing	30−34	
Dividing wall, 6−12 cm brick	37−42	
Dividing wall, 25−38 cm brick	50−55	
Double wall, 2×12 cm brick	60−65	

Tackling noise at source

The most effective and rational way to deal with noise is usually to tackle it at source. Certain jobs, and certain pieces of machinery, are noisy because heavy, hard surfaces clash together. Sometimes this noise can be reduced by replacing the hard material with something softer, e.g., vulcanized rubber, ordinary rubber or felt. For the same reason transport vehicles are quieter with rubber tyres than with metal wheels. All forms of transmission should be examined to see if they are still in good condition, since old and worn components create unnecessary noise. Belt drives, using rubber, leather or fabric belts, are quieter than a series of cog-wheels. Toothed belts engaging with toothed wheels are the quietest of all, especially if the toothed wheels are made of some synthetic material.

Noise radiated from vibrating plates can be reduced by stiffening them, loading them with weights, making them curved, or using non-resonant materials. Moving machinery, and motors in operation not only send out sound waves, but also set the structure of the building in vibration. Such vibrations and resonances, and the secondary noises that they set up, can be disturbing throughout the whole building. For this reason very heavy machines should be rigidly set in concrete or iron mountings. If necessary they can be mounted in special concrete troughs, with intervening layers of sound-insulating material. Such sound-insulating layers can be of spring steel, rubber, felt or cork, according to the weight of the offending machine.

Enclosing the source of noise

An especially effective way of reducing noise is to enclose the source. A housing of suitable material may reduce the radiated noise by 20–30 dB. The inside wall of such a housing should be lined with sound-absorbing material, while the wall itself should be as heavy and as airtight as possible. The housing should enclose the source of noise, with as few gaps as possible. Of course some apertures are usually necessary for the passage of leads, or for operating the machine; such apertures impair the soundproofing of the machine, and it can be stated as a general guideline that their total area should not exceed 10% of the area of the housing. Doors or sliding panels may be provided to allow the machine to be controlled and serviced.

Sound insulation in a room

When all the possibilities of sound damping at source and by enclosure have been exhausted, it is sometimes possible, in the right circumstances, to give further protection by covering the walls and ceilings with sound-absorbing material. *Sound-absorbing panels (acoustic tiles) absorb part of the sound, and so reduce reflection back into the room and the echo effect.*

Up to the present, acoustic tiles have been used mainly in offices of more than 50 m^2 in area, in accounts offices, in cashiers' offices, and in offices giving counter services. Sound reductions of 5 dB, or even up to 10 dB can be expected in such places. On the other hand, the effects of installing acoustic tiles in noisy workshops, machine rooms and factories are not always clear or easy to assess. It must be remembered that when an operative works close to a source of noise he is mainly affected by the sound reaching him direct, and reflected sound is of little importance to him. Covering the walls and ceiling with acoustic tiles will therefore give no protection to such a worker. Ceiling tiles will be of slight benefit to him only if he is working several metres from the source of the noise.

Personal ear protectors

If an operative is forced to work in a noisy environment, and technological means have failed to reduce the noise to a safe level of below $L_{eq}=85$ dB(A), then as a last resort there remains only personal ear protection.

The following possibilities are worth considering:

1. *Ear-plugs to block the outer ear passages; these may be of cotton-wool, wax or synthetic material.*
2. *Protective caps, which cover the entire external ear (ear-muffs).*
3. *A sound-proof helmet, enclosing the head, including the ears.*

Ear plugs

A simple plug of cotton-wool or wax in each outer ear passage is an old device that is often used. *Properly used, ear-plugs can reduce the noise level by up to 30 dB.* Instead of cotton-wool, conical plugs of a synthetic material (Selectone) are available, which cut out more of the higher frequencies than the lower ones, and so interfere less with conversation. All types of ear-plugs have the drawback that they need to be used carefully and correctly if they are to give sufficient protection, and if used regularly they may set up irritations, such as aural eczema.

Protective caps (ear-muffs)

Caps which enclose the whole of the external ear (ear-muffs) give good protection, and if correctly fitted, can reduce noise levels by about 40−50 dB in the frequencies 1000−8000 Hz.

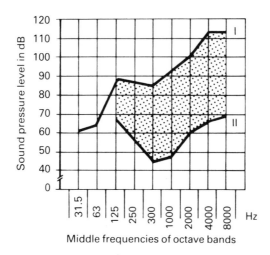

Figure 171. **Protective effect of an ear-muff in the vicinity of a circular saw cutting light metal.**
I = noise spectrum at the unprotected ear. II = noise spectrum inside the ear-muff.

The insulating pad must fit closely round the ear, on to the skull, so that it does not exert any pressure on the ear itself, otherwise it may set up pains and headaches. Figure 171 shows the effects of using an ear-muff. The graph shows that the noise of a circular saw, with frequencies between 500 and 8000 Hz, is reduced at the eardrum by about 40−45 dB.

Unfortunately many workers object to any kind of ear protectors. Their main ground for complaint is 'acoustic isolation', the worker thinks that he or she is missing some vital information from the surroundings. He or she also thinks that output, and hence earnings, will be adversely affected. In spite of these objections, the wearing of ear protectors should be encouraged in the interests of the workers themselves.

As general guidelines we might suggest ear plugs for noise levels 85−100 dB(A); and ear-muffs for noise levels above 100 dB(A).

Medical tests

Noise protection in its widest sense also includes those medical measures that are laid down for noisy places of employment in many countries, and which provide for periodic audiometric tests. The purpose of these is:

1. To detect noise damage at its onset.
2. To supply a basis for devising protective measures against noise, and for the introduction of personal protection into factories.

19.6. Vibrations

Vibrations are mechanical oscillations, produced by either regular or irregular periodic movements of a body about its resting position. They are termed mechanical oscillations because in the last analysis it is the small changes of position that are important.

The nature of vibrations, and their effects on people have been written about in detail by Wisner [276], Grether [95], in the ISO Publication [129] and in the Guidelines of the British Standards Institute [27].

A little physics

The following five physical quantities are important for the understanding of what follows.

1. *Point of application to the body.* Two points at which vibrations enter the body are significant ergonomically: feet/buttocks (when driving or riding in a vehicle) and hands (when operating vibrating tools or a machine). The direction of oscillation is important. Most often this lies in the vertical plane (head to foot) or approximately along the line of hand and arm.

2. *Frequency of oscillations.* The extent of the physiological and pathological effects of vibrations is strongly frequency dependent. Particularly important frequencies are those which fall into the range of natural frequencies of the body, and so cause resonance (see item no. 5). Often a low and high range of frequencies are distinguished. The threshold lies between 30 and 50 Hz. Vibrations of motor vehicles belong to the low, those of motor-driven tools to the high ranges of frequencies.

3. *Acceleration of oscillations.* Within the frequency range that is physiologically important, *the acceleration of the oscillations is usually taken as a measure of the vibrational load. The unit is the acceleration due to gravity* $g = 9 \cdot 8 \text{ m/s}^2$.

4. *Duration of effect.* The effect of vibrations depends greatly on their duration. Their ill-effects increase very rapidly as time goes on.

5. *Individual frequency and resonance.* Every mechanical system which possesses the elementary properties of mass and elasticity is capable of being set in oscillation. The force which sets the system in motion is known as the *inducing* force, and the resultant oscillations are called *the forced vibrations.*

Each system has its own natural frequency, and the nearer the frequency of the inducing force comes to this, the greater will be the amplitude of the forced vibrations. When the amplitude of the forced vibrations exceeds that of the inducing force the system is said to be *in resonance.*

On the other hand the oscillations of any system are subject to *damping*, which reduces their amplitude. Thus, for example, when we are standing up any vertical vibrations set up in the legs are quickly dampened.

Frequencies above 30 Hz are particularly heavily dampened by the tissues of the body; thus for an inducing frequency of 35 Hz the amplitude of oscillation is reduced to 1/2 in the hands, to 1/3 in the elbows and to 1/10 in the shoulders.

Oscillatory characteristics of the human body

The human body does not vibrate as a simple mass, with its own natural frequency. Studies have shown that the natural frequencies are different in different parts. The body of a sitting person reacts to vertical vibrations as follows:

3−4 Hz: strong resonance in the cervical vertebrae.
4 Hz: peak of resonance in the lumbar vertebrae.
5 Hz: very strong resonance in the shoulder girdle (up to 2-fold increase).
20−30 Hz: resonance between head and shoulders.
60−90 Hz: resonance in the eyeballs.
100−200 Hz: resonance in the lower jaw.

In general, the most effective inducing frequency for vertical vibrations lies between 4 and 8 Hz. In more detail:

1. *Vibrations between 2·5 and 5 Hz generate strong resonance in the vertebrae of the neck and the lumbar region* (oscillations amplified up to 240%).
2. *Between 4 and 6 Hz resonances are set up in the trunk, shoulders and neck* (amplification up to 200%).
3. *Between 20 and 30 Hz sets up the strongest resonances between the head and shoulders* (amplification up to 350%).

There has been little study of other points of application, and oscillations in other directions. It can only be said that the natural frequencies of the smaller components of the body, such as groups of muscles, eyes, and so on, lie in the higher frequency ranges. Hence the operation of machines with frequencies above 30 Hz are likely to set up resonances in the fingers, hands and arms. On the other hand the damping effect of bodily tissues is greater for these higher frequencies, and tends to confine them to the vicinity of the point of application.

Vibrations at the workplace

Up to the present, vibration experienced at work has been measured mainly in construction machinery, tractors and trucks. The studies on various motor vehicles [62] revealed that *the acceleration of vertical oscillations lies between 0·5 and 5 m/s^2*, with the highest values being recorded for earth-moving machines and tractors. Operating motor-powered tools involves high levels of vibration in the hands and wrists, some examples of which are summarized in Table 42.

Table 42. Vibration in motorized hand-tools.
Effective acceleration = root mean square of the accelerations at various amplitudes. These figures apply to oscillations along the direction of the arm; oscillations perpendicular to this are often higher. After Dupuis [62].

Type of tool	Effective acceleration (m/s^2)		Ground frequency (Hz)
	On fingers	On wrist	
Power saw	17·5	1·1	120
Soil borer	21·0	3·5	110
Pneumatic compass saw	—	9·9	—
Two-wheeled cultivator	3	2·8	82

Physiological effects

Vibration seriously affects visual perception and psychomotor performance and the musculature, circulation and respiratory system to a lesser extent.

Vibration seems to generate muscle reflexes which have a protective function, causing the extended muscle to shorten. The reflex activity of the muscles also explains the often-observed increase in energy consumption, heart rate and respiratory rate when a person is exposed to strong vibrations. These vibrational effects on metabolism, circulation and respiration are small and have little significance.

Visual powers

The adverse effect of vibration on eyesight is most important because it impairs the efficiency of drivers of tractors, trucks, constructional machines and other vehicles, and increases the risk of accidents. *Visual acuity is poorer* [97], *and the image in the visual field becomes blurred and unsteady* [96].

Visual powers are not affected by vibrations of less than 2 Hz. Measurable optical aberrations appear from 4 Hz upwards, and are greatest in the range 10−30 Hz. With a vibration of 50 Hz and an oscillatory acceleration of 2 m/s^2, visual acuity is reduced by half, according to Guignard [97].

Skill

Strong vibration impairs performance in various psychomotor tests. Simplifying somewhat, we may say that strong vibration impairs visual perception, mental processing of information, and the carrying out of skilled movements [95, 97].

Driving tests

These psychophysiological effects of vibration are particularly evident in all kinds of simulated driving tests:

1. Over the range 2−16 Hz (especially around 4 Hz) driving efficiency is impaired, and the effects increase with increasing acceleration of the oscillations.
2. Driving errors increase when the seat is subjected to accelerations of the order of 0·5 m/s^2.
3. When the accelerations reach 2·5 m/s^2, the number of errors becomes so great that such vibrations must be rated as positively dangerous.

This consensus of physiological effects of vibration point to the conclusion that *strong mechanical oscillations reduce efficiency, and in many situations may lead to the risk of errors and accidents.*

Vibration as a nuisance

Vibration is subjectively felt as an imposition and a burden, impressions ranging from a minor annoyance to an unbearable nuisance. The extent of the nuisance depends in the first instance on the inducing frequency, on the rate of acceleration of the oscillations, and on the length of time they continue. The source of the nuisance lies in the physiological effects, and in the resonances set up in various parts of the body. Figure 172 shows the results of an investigation by Chaney [43] on seated test subjects; curves are given for subjective feelings of

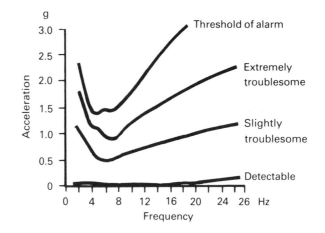

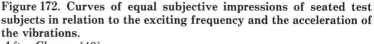

Figure 172. **Curves of equal subjective impressions of seated test subjects in relation to the exciting frequency and the acceleration of the vibrations.**
After Chaney [43].

equal intensity, in relation to frequency and acceleration of oscillation.

From these results the following conclusions may be drawn for vertical oscillations applied to seated people.

1. *The most intense subjective sensitivity lies in the frequency range 4—8 Hz.*
2. *The average threshold of 'very severe' intensity comes at an acceleration of 1 g (i.e., about 10 m/s²).*
3. *At accelerations of 1·5 g (i.e., about 15 m/s²) the vibrations became dangerous and intolerable.*

The same author carried out similar tests on standing subjects. The curves of equal subjective intensity were higher than before, because of the damping effects in the legs, to which we referred earlier. The threshold of 'very severe' intensity came 0·2—0·3 g higher than for seated subjects.

Complaints

The complaints which are suffered in addition to the annoyance of vibration vary greatly, but some of them are frequency-dependent. The following are the most common complaints:

1. Interference with breathing, especially severe at vibrations of 1—4 Hz.
2. Pains in chest and abdomen, muscular reactions, rattling of the jaws, and severe discomfort, chiefly at 4—10 Hz.
3. Backache, particularly at 8—12 Hz.
4. Muscular tension, headaches, eyestrain, pains in the throat, disturbance of speech, irritation in intestines and in the bladder, at frequencies of 10—20 Hz.

In addition we may mention sea and travel sickness, with

nausea and vomiting, brought about by oscillations of $0 \cdot 2 - 0 \cdot 7$ Hz, with the greatest effect at $0 \cdot 3$ Hz.

Damage to health

Exposure to vibration at one's place of work can, if repeated daily, lead to morbid changes in the organs affected. The effects are different in the two parts of the body most commonly subjected to vibration. Vertical oscillations when either standing up or sitting down, caused by vibration from underneath (e.g., in a vehicle) can cause degenerative changes in the spine, whereas the vibrations of power tools affect mainly the hands and arms.

Spinal ailments

Tractor drivers in various countries have been recorded as suffering an accumulation of disc troubles and arthritic complaints in the spine, as well as an above-average incidence of intestinal ailments, prostate troubles and haemorrhoids [97, 62].

The cumulative appearance of spinal damage among workers who have been subjected to a high level of vertical oscillation leads one to suppose that heavy and prolonged vibration causes excessive wear and tear on the intervertebral discs and the joints. This is still hypothetical, however, since the causal link has not yet been firmly established.

Hand and arm troubles

Workers who use power tools for years on end (e.g. chain-saws or pneumatic hammers) may suffer various ailments of the hands and arms, and, according to Wisner [276] the frequency of vibration is a decisive factor.

Arthritis

Tools with a frequency vibration below 40 Hz, e.g., a heavy pneumatic hammer, can cause degenerative symptoms in the bones, joints and tendons of hands and arms, leading to arthritis in the wrist, elbow and occasionally in the shoulder.

Atrophy

The effects on the bones may lead to atrophy, which in rare cases may involve so much loss of calcium that the risk of fracture is substantially increased. In a few countries these possible consequences of using a pneumatic hammer are classified as an occupational disease.

'Dead fingers'

Power tools with frequencies between 40 and 300 Hz usually have a very small amplitude of oscillation ($0 \cdot 2 - 5$ mm), and their vibrations are quickly dampened in the tissues. *Such vibrations may have ill-effects on the blood vessels and nerves of the hands*, resulting in one or more of the fingers going 'dead'. Usually the middle finger is most affected, becoming white or bluish, cold and numb. After a little while the finger turns pink again and is painful. The cause is a cramp-like condition of the blood vessels, known as Raynaud's disease.

These symptoms have also been observed among miners

using a pneumatic drill with a higher frequency of vibration, as well as among forestry workers using power saws with frequencies between 50 and 200 Hz.

'Dead fingers' make their appearance, at the earliest, six months after the beginning of the work using the vibrating tool, and cold is an important factor in the onset of this condition. Raynaud's disease is more common in northern countries than in warmer latitudes. It must be assumed that cold makes the blood vessels more sensitive to vibration, and more liable to vascular cramps.

Users of power tools with even higher frequencies have also been known to suffer from ailments affecting the circulation, and loss of sensation. One example of such tools is polishing machines operating at 300−1000 Hz, which caused painful swellings and loss of sensation in the hands, which often did not disappear when the work was finished.

Limiting values   Attention has been given in many countries to the question of fixing limits for what may or may not be expected in the way of vertical oscillations [62, 97].

From these studies it may be concluded that vibration becomes intolerable in the following circumstances:

Below 2 Hz: at accelerations of $3-4\ g$
Between 4 and 14 Hz: at accelerations of $1\cdot2-3\cdot2\ g$
Above 14 Hz: at accelerations of $5-9\ g$.

The International Standards Organization (ISO) has been concerned with this problem for several years; its guidelines for the assessment of vertical oscillations are shown in simplified form in Figure 173.

This ISO Standard distinguishes between three criteria, and envisages three levels of limiting values:

1. *The criterion of comfort* (reduced comfort boundary). This applies mostly to vehicles, and to the automobile industry.
2. *The criterion of the maintenance of efficiency* (fatigue-decreased proficiency boundary). A decisive factor in this criterion is working efficiency (proficiency). It applies to tractors, construction machinery and heavy vehicles. These limiting values are shown in Figure 173.
3. *The criterion of safety* (exposure limits). Protection against damage to health is the criterion here.

Starting from the 'efficiency' of the second criterion (Figure 173), the other two can be derived as follows:

The criterion of comfort can be derived by dividing the acceleration by 3·15.
The criterion of safety can be derived by multiplying the acceleration by 2.

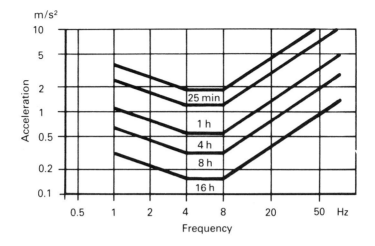

Figure 173. Threshold values for vertical vibrations before efficiency is affected.
'Fatigue-decreased proficiency boundary', for exposure times ranging from 25 minutes to 16 hours. For the calculation of the threshold values 'safety' and 'comfort', see text. Note exponential scales. After ISO [129].

Recommendations

From the standpoint of ergonomics, tractors, heavy vehicles and construction machinery, with their frequencies most often between 2 and 5 Hz and operating for an 8-hour day, require a limit of oscillational acceleration of $0 \cdot 3 - 0 \cdot 45$ m/s^2. These limits are often exceeded.

Driving seats

It is clear from this that *the ergonomic construction of driving seats requires them to be so dampened that vertical oscillations are reduced to within these safety limits.*

It is already possible to do this by special types of suspension. By 1956 Simons *et al.* [242] had already developed driving seats in which the impact of vibrations in the frequency range 3—5 Hz had been reduced by more than 50% Dupuis [62] obtained similar results with special seats which reduced the effects of vibration by 50—60%.

Damping hand-operated power tools

The same applies in principle to hand-operated power tools. Damping elements within the tool itself, and even more between the tool and the hand-grip, can reduce vibration considerably. The grip itself can be dampened by making it of flexible material. Further improvement can be obtained by wearing thick gloves and avoiding working in very cold conditions.

20. *Indoor climate*

20.1. Thermoregulation in man

The indoor climate, as understood in this Chapter, means the physical conditions under which work is carried on. The principal components are:

1. Air temperature;
2. The temperature of the surrounding surfaces.
3. Air humidity;
4. Air movements.
5. Air quality.

Body temperature

The temperature of the human body is not, as often assumed, uniform throughout. A constant temperature, which fluctuates a little around 37°C, is found only in the interior of the brain, in the heart and in the abdominal organs (core temperature). A constant core temperature is a prerequisite for the normal functioning of the most important vital functions, and wide or prolonged deviations are incompatible with the life of a warm-blooded animal. In contrast to the core temperature, that in the muscles, limbs, and above all in the skin (the shell temperature) shows certain variations. Physiological studies have shown that when the surrounding air is cool there is a steep temperature gradient in the skin, from the inside outwards. For example, in cool air, the temperature even 20 mm beneath the surface of the skin may have fallen to 35°C, whereas in warmer surroundings it may still be 35–36°C only a few millimetres below the surface. This capacity for adaptation allows the human body to tolerate a temporary heat deficit which may be in the region of 1000 kJ for the whole body. The muscles too, show considerable fluctuations in temperature, being several degrees warmer during strenuous effort than when they are at rest.

Control processes

The control mechanisms throughout the body, which are necessary to maintain a constant core temperature, are shown diagrammatically in Figure 174. At the centre of the control system is the heat control centre located in the brain stem.

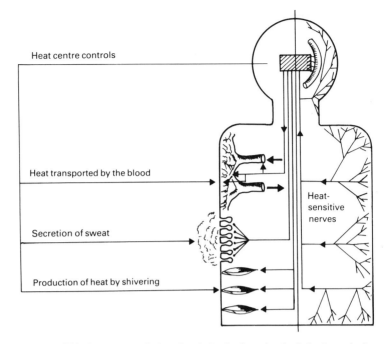

Heat centre controls

Heat transported by the blood

Heat-
sensitive
nerves

Secretion of sweat

Production of heat by shivering

Figure 174. Diagram of the physiological control of the heat balance of the body.
The heat centre located in the brain stem regulates the flow of blood through the capillaries of the skin, as well as the secretion of sweat. These two mechanisms, between them, adjust the heat balance of the body according to external and internal conditions.

This is the overall director of the thermal regulation of the body, and so is comparable to a room thermostat.

The nerve cells of the heat control centre receive information about temperatures throughout the body, sometimes directly and sometimes from heat-sensitive nerves in the skin. The heat control centre in turn sends out the impulses that are necessary to direct and control the regulatory mechanism which keeps the core temperature constant. In this way the heat production of the body, its diffusion by way of the circulatory system, and the heat loss by secretion of sweat in the skin are controlled, thus enabling the process of thermoregulation to be carried on.

Heat transport by the blood

The most important item in thermoregulation is the heat transport function of the blood, whereby the blood vessels, especially the capillaries, act as distributors of heat, picking up heat from warm tissues and giving it out again to cooler tissues. In this way the blood can transport heat from the interior of the body to areas of skin that are cooled by the external temperature; conversely, if the exterior is artificially heated, heat can be transported away into the interior. *The key to this mechanism is the control of blood circulation in the skin.*

Secretion of sweat	The second regulatory mechanism directed by the heat control centre is the secretion of sweat in the skin. This, too, is under nervous control.
Shivering	The third regulatory mechanism is to raise the rate at which heat is produced by the body, a process that is set in motion whenever the body is subjected to cooling. This occurs by an increase in metabolic heat in the muscles and other organs, a special manifestation of which is the rapid muscular movements known as 'shivering'.
Heat exchange	As explained earlier, the body converts chemical energy into mechanical energy and heat. The body uses this heat to maintain a constant core temperature and dissipates any excess heat into its surroundings. There is thus a constant exchange of heat between the body and its surroundings which is regulated in part by physiological control, and partly by the ordinary laws of physics. The latter involve four different processes: 1. Conduction. 2. Convection. 3. Evaporation. 4. Radiation.
Conduction of heat	Heat exchange by conduction depends first and foremost on the conductivity of objects and materials in contact with the skin. Anyone who sits down in winter, first on a stone and then on a tree-trunk can discover this for himself. The stone feels very cold because it conducts heat away from the body; the tree-trunk feels much less cold because its conductivity is less. Conductivity of heat is of practical importance in the choice of floorings, furniture and the parts of machinery that have to be touched (control handles, etc.). Loss of heat by contact with such objects is to be avoided, either from the feet or from other parts of the body; it is decidedly unpleasant, and is liable to cause ailments such as rheumatism and arthritis. Hence *workplaces should have well-insulated floors (e.g., cork, linoleum or wood), while bench tops, machine parts (particularly handles and controls), and tools should be protected with felt, leather, wood or other material of low thermal conductivity.*
Convection	Heat exchange by convection depends primarily on the difference in temperature between the skin and the surrounding air, and on the extent of air movement. Under normal circumstances convection accounts for about 25–30% of the total heat exchange of the body.

Evaporation of sweat

Loss of heat by sweating occurs because the sweat on the skin evaporates, consuming heat. This latent heat of vaporization amounts to $2 \cdot 3$ kJ/g water evaporated. Under normal conditions each person will evaporate about one litre per day (insensible perspiration), and lose 2500 kJ or more of heat in the process: about one-quarter of the total daily loss of heat.

If, however, the temperature of the surroundings exceeds the limits of comfort, then the hot skin begins reflex sweating, with a sharp increase in the rate of loss of heat.

The extent of heat loss by the evaporation of water depends on the area of skin from which sweat can evaporate, and on the difference in water vapour pressure between the air next to the skin and that further away. Thus *the relative humidity of the surrounding air is an important factor*. A lesser factor is that of air movement, which on the one hand increases the gradient of water vapour pressure, but on the other hand cools the skin by convection, and thereby reduces the amount of sweating.

At environmental temperatures (of walls and of the air) above 25 °C the clothed human body is able to lose hardly any heat by either convection or radiation, and sweating is the only compensatory mechanism left. Hence the loss of heat by evaporation of sweat rises steeply after a particular critical temperature has been reached.

Radiation of heat

Warm bodies radiate electromagnetic waves of relatively long wavelength, which are absorbed by other bodies (objects and surfaces) and converted into heat. This is called infra-red radiation, or radiant heat. It does not depend on any material medium for its transmission, in contrast to conducted or converted heat. Such heat exchange by radiation takes place between the human body and its surroundings (walls, inanimate objects, other people) in both directions, and all the time. In contrast to conduction or convection, heat radiation is hardly affected by the temperature, humidity or movement of the air; it depends mainly on the temperature difference between the skin and adjacent surfaces. In temperate countries the surrounding objects and surfaces are usually cooler than the human skin, and so the human body loses a considerable amount of radiant heat in the course of a day.

Loss of heat by radiation is not noticeable as long as the amount is not excessive, but it may become uncomfortable when standing close to a cold wall or a large window, even though the air temperature is high enough. In such circumstances the loss of heat may be considerable, because the decisive factor is not the air temperature, but the temperature difference between the skin and the adjacent cold surface.

The amount of radiant heat loss in a day by a fully clothed person varies very greatly, according to circumstances. Excluding high summer, the average daily loss in temperate climates is about 40−60% of the total heat lost from the body.

Figure 175 shows three of the principal ways in which the
human body exchanges heat with its surroundings.

To summarize, it can be said that the following four physical
factors are decisive:

1. Air temperature for exchange of heat by convection.
2. Air movement, also for convection.
3. Temperatures of adjacent surfaces: walls, ceiling, floor,
   machinery, etc. for exchange of heat by radiation.
4. Relative humidity of the air for loss of heat by evaporation
   of sweat.

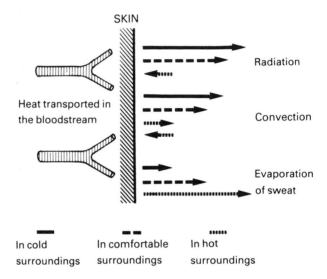

**Figure 175. Diagram of heat exchange between the human body and
its surroundings.**
*The length of the arrows gives a rough indication of the heat
transferred by each of the three processes, under different conditions.*

20.2. Comfort

*Physiological
basis of comfort*
One hardly notices the internal climate of a room as long as it
is comfortable, but the more it deviates from a comfortable
standard, the more it attracts attention.

The sensation of discomfort can increase from mere annoyance to pain, according to the extent to which the heat balance
is disturbed. Discomfort is a practical biological device in all
warm-blooded animals, which stimulates them to take the
necessary steps to restore a correct heat balance. An animal
can only react by seeking out another place that is neither too
hot, nor too cold, but man has the use of clothing, as well as
being able to modify his environment by technological means.

*Side-effects of
discomfort*
Discomfort brings about functional changes that may affect
the entire body. Overheating leads to weariness and sleep-

iness, loss of performance and increased liability to errors. This damping down of activity causes the body to produce less heat internally.

Conversely, overcooling induces restlessness, which in turn reduces alertness and concentration, particularly on mental tasks. In this case stimulation of the body into greater activity causes it to produce more internal heat.

Thus the maintenance of a comfortable climate indoors is essential for well-being and maximum efficiency.

Temperature zones in physiological terms

If a test subject is placed in a climatic chamber and exposed to different temperatures, a range can be found in which the heat exchanges of his or her body are in a state of balance. This is called the *zone of vasomotor regulation*, or the *comfort zone*, because within this range the heat balance is maintained chiefly by regulating the flow of blood to different parts of the body. For a clothed and resting person in winter this zone lies between 20 and 23 °C.

If there is a slight excess of heat above this comfort level, this is dealt with by warming the peripheral parts of the body and by increasing perspiration. This is the *zone of evaporative control*. If, however, the heat continues to increase and exceeds a certain level, *the limit of tolerance*, the core temperature rises quickly and steeply, and in quite a short time leads to death by heat stroke.

Temperatures below the zone of vasomotor regulation are characterized by a negative heat balance for the body, since more heat is being lost than is being generated internally. This is the *zone of bodily cooling*. At first the cooling is confined to

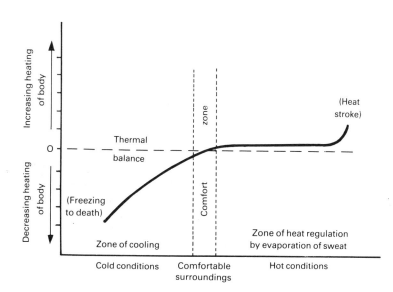

Figure 176. Heat balance of the body between exposure to extremes of heat and cold.

the peripheral parts of the body, which can tolerate a heat deficit for a time.

The heat balance in these three zones is shown diagrammatically in Figure 176.

Comfortable ranges of temperature

If test subjects are asked to say when they feel really comfortable, the range is a comparatively narrow one, perhaps only 2 or 3°C. Obviously people feel comfortable only when the vasomotor regulation system is not heavily stressed, that is when the blood circulation to the skin is subject to no more than normal fluctuations. On the other hand either a negative or a positive heat balance (i.e., either a deficit or an accumulation of heat in the body shell) is felt as discomfort.

The temperature range within which a person feels comfortable is very variable. It depends firstly on the amount of clothing being worn, and then on how much physical effort is being performed. Whether other factors such as food, time of year, time of day, age and sex are equally important is arguable.

Four climatic factors and comfort

People's impressions of comfort are influenced by the same four climatic factors that determine the heat exchange, so we may repeat them here:

1. *Air temperature.*
2. *The temperatures of adjacent surfaces.*
3. *Air humidity.*
4. *Air movements.*

Thus each factor is involved in its own 'balance', and several research workers have tried to find a unit of measurement that would take account of them all. An example is the so-called 'Kata-value' which attempts to use the rate of cooling of an artificial body as an index of comfort, but this method has not been put into practice. Research workers have come to rely more and more on the subjective impressions of test subjects as a measure of the degree of comfort, resulting in the concept of *effective temperatures.*

Effective temperatures

Houghten and Yaglou [120], in 1923, were the first to investigate the relationship between temperature and humidity of the air that would result in the same effective temperature. The test subjects were placed in a climatic chamber in still air, at a given temperature, and with 100% relative humidity. They were required to note their impressions of the temperature, and to remember them. Then the relative humidity was reduced, and the temperature varied until the test subjects felt the same sensation of warmth as before. This is the *effective (or perceived) temperature.* These investigations were followed by others, in which the effects of temperatures of adjacent surfaces and of air movements on the effective temperature were analysed. These led to the formulation of 'indices of comfort', and 'zones of comfort', which could be read off from nomograms in relation to three or four of the climatic factors listed above. We may refer to the experiments of the Danish worker Fanger [70], who compiled an equation of comfort out of the ratio between the four climatic factors, the amount of clothing, and the extent of physical activity. This equation is very complex, however, and results can be worked out only by using a computer. Fanger [70] set out the most important results in 28 diagrams, showing curves of degree of comfort in relation to several factors. An account of these would be beyond the scope of this Chapter, so we must confine ourselves to considering a few relationships between the climatic factors already mentioned which are important in the evaluation of the climate indoors. The leading question is what is the effect on effective temperatures of various combinations of:

air temperature and temperature of surrounding surfaces;
air temperature and relative humidity; and
air temperature and air movement?

Air temperature and that of adjacent surfaces

Physiological research has shown that the effective (perceived) temperature is essentially a mean between that of the air and of adjoining surfaces. Expressed as a formula:

$$\text{Effective temperature} = \frac{T_A + T_S}{2}$$

where T_A = mean air temperature, and T_S = mean temperature of adjacent surfaces.

It is important for comfort that the difference between T_A and T_S should be small. Large areas of cold walls or windows are particularly uncomfortable, even if the air temperature is adequate. *It is a good rule of thumb that the mean temperature of adjacent areas should not differ from that of the air by more than 2 or 3°C, either up or down.*

Temperature and humidity of the air

The effect of atmospheric humidity was prominent in early research [120], but more recent work [145, 201] has shown that after a prolonged stay in the same room, the impression of temperature is little affected by the humidity of the air. The following combinations of atmospheric humidity (%) and air temperature will produce equal effective temperatures:

70% RH and 20°C
50% RH and 20·5°C
30% RH and 21°C

So it is clear that within the range 30−70%, relative humidity has little influence on effective temperature. It can be assumed that at 18−24°C the relative humidity can fluctuate between 30 and 70% without creating thermal discomfort. The threshold at which the room begins to feel stuffy lies between the following pairs of values:

80% RH and 18°C
60% RH and 24°C

If the relative humidity falls below 30% there is a danger of too dry an atmosphere, about which we shall speak briefly later on.

Temperature and movement of the air

The third factor to influence effective temperature is air movement. Yaglou [280], carried out experiments to determine the ratios of air temperature and free movement of the air, which combined to give the same effective temperature of 20°C. His results are set out in Table 43.

More recent research [70] has shown that air movements in excess of 0·5 m/s are unpleasant even when the air is warm, and that the discomfort depends on the direction in which the air is flowing, and on the parts of the body exposed to it.

1. *Air currents from behind are more unpleasant than those coming from in front.*
2. *Neck and feet are particularly sensitive to draughts.*
3. *A cool draught is more unpleasant than a warm current.*

From general experience it can be said that seated people find draughts of more than 0·2 m/s unpleasant. Occasionally, when very precise work is being carried out, which keeps the operator motionless for long periods, even draughts of 0·1 m/s can be unpleasant. In contrast, standing work, especially if it involves strenuous physical effort, can be carried out without ill-effects in draughts of up to 0·5 m/s.

Table 43. **Air movement and effective temperature.**
After Yaglou [281]. The two columns show the dry-bulb and wet-bulb temperatures of a psychrometer (wet- and dry-bulb hygrometer).

Air movement (m/s)	Dry bulb (°C)	Wet bulb (°C:100% RH)
0	20·0	20·3
0·5	21·0	21·3
1·0	22·0	22·2
1·5	22·8	23·0
2·0	23·5	23·8

20.3. Dryness of the air

As we have seen, dryness of the air is a winter problem both at work and at home, hygienically and medically. For many years the tendency has been to prefer higher and higher temperature indoors during the period of the year when buildings are heated, and this has led to lower and lower values of relative humidity.

In one of our investigations we carried out random tests in a number of offices, both in winter and in summer, to measure the relative humidity of the air. The results are shown in Figure 177.

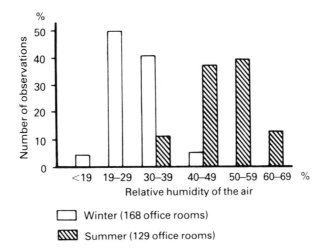

Figure 177. **Relative humidity of the air in summer and in winter.** *Winter samples taken in 168 offices without air conditioning. Summer samples taken in 60 offices with and 69 without air conditioning.*

Medical effects

Most ear, nose and throat specialists think that the present-day tendency to excessive dryness of the air in heated rooms causes an increased incidence of catarrhal ailments, and chronic irritation of the nasal and bronchial passages. They often note a desiccation of the mucous membranes of the air ducts, which they think obstructs the flow of mucus over the ciliary tracts, resulting in a diminished resistance to infection. These views of the effects of dry air are shown diagrammatically in Figure 178.

These observations may be summarized by saying that relative humidities of 40–50% in heated rooms are desirable for comfort and below 30% become unhygienic, because they adversely affect the mucous membranes of nose and throat.

It may be mentioned here that most of the humidifiers sold commercially are inadequate for their purpose. As a rough guide, a workroom of 100 m^3 volume requires a minimum water output of 1 l/h.

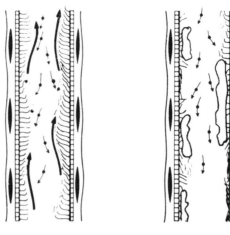

Normal air-passage

Dust particles adhere to the
mucus and are carried away
with it by the ciliated epithelium

Desiccated air-passage

Mucus has coagulated into
clumps to which dust particles
do not adhere

Figure 178. Diagram showing the effects of dry air on the self-cleaning capacity of the mucous membranes in the throat.
Left: normal mucous membrane with a ciliated epithelium which rids the membrane of any contaminating particles.
Right: the ciliated epithelium is desiccated, and parts of it are no longer visible—the mucus forms into clumps and does not get rid of the dust particles.

20.4. Field studies on indoor climate

We have already seen that there is considerable individual variation in what people consider to be a comfortable temperature, and field studies emphasize this. Figure 179 shows the results of some early research by McConnell and Spiegelman [181], which recorded the views of 745 office workers on temperatures in New York during the summer of 1940.

The most striking result was the wide range of individual responses. At best (24°C) only 65% of those questioned found this temperature comfortable, all the rest finding it either too hot or too cold. In one of our own studies [84], carried out in the winter of 1964/65, we took measurements in 168 offices without air conditioning and at the same time questioned 410 employees (140 men and 270 women) about their impressions of temperature. The results are shown graphically in Figure 180.

Air temperatures were high, varying between 22°C and 24°C most of the time. The reply "the room is pleasant" became less frequent as the temperature rose, while "too warm" took its place. *There were many replies of "pleasant" when the temperature was only 21°C.*

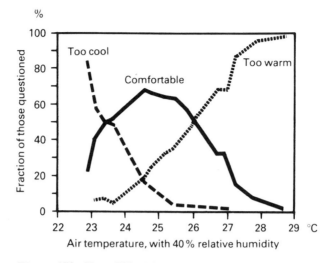

Figure 179. How 745 office workers in New York in the summer of 1940 reacted to the air temperature.
After McConnell and Spiegelman [181].

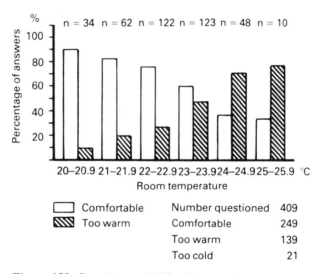

Figure 180. Reactions of 409 office workers to temperature in the winter 1964/5.
n = number of people questioned and also of temperature records. The question asked was: 'Do you find the indoor climate in the office today comfortable, too warm or too cold?' After Grandjean [84].

Similar studies were carried out in the summers of 1966 and 1967. This survey comprised 311 offices, 122 of them air-conditioned, and 189 not. A total of 1191 employees were questioned, and the results are summarized in Figure 181.

In the offices without air conditioning, the air temperatures mostly ranged between 20°C and 27°C; in the air-conditioned offices the upper limit of temperature was usually 24°C. As

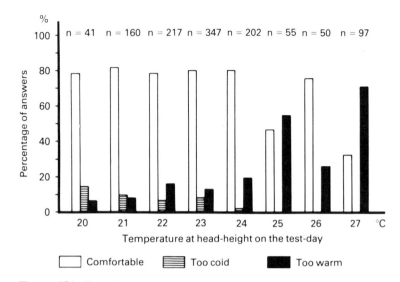

Figure 181. Reaction to temperatures in the summer.
*n = number of people questioned, and also of temperature readings.
The question asked was: "Do you find the indoor climate in the office
today comfortable, too warm or too cold?" After Grandjean [84].*

can be seen from the diagram, temperatures above 24°C were
assessed as "too warm" by most people, and a similar observation was made in the US Survey of McConnell and Spiegelman
[181]. So we may conclude that *in summer room temperatures
are comfortable as long as they do not exceed 24°C.*

20.5. Recommendations for comfort indoors

*Sedentary office
work*

As far as present knowledge goes, the following guidelines
may be applied to sedentary work which does not involve
manual effort:

1. *The air temperature* in winter should be between 20 and
   21°C and in summer between 20°C and 24°C.
2. *Surface temperatures of adjacent objects* should be at
   roughly the same temperature as the air or at least not
   more than 2–3°C different. No single surface (e.g., the
   outside wall of the room) should be more than 4°C colder
   than the air in the room.
3. *The relative humidity of the air in the room* should not fall
   below 30% in winter, otherwise there will be a danger of
   desiccation problems in the respiratory tract. In summer
   the natural relative humidity usually fluctuates between
   40% and 60% and is considered comfortable.
4. *Draughts* at the levels of head and knees should not exceed
   0·2 m/s.

It must be pointed out that the preferred air temperatures might deviate slightly from one country to another, mainly because of clothing differences and also for traditional reasons. Thus it is well known that room air temperatures are generally higher in the USA and often lower in the UK. But the above guidelines are certainly justifiable from a physiological point of view.

Air temperatures for physical work

With physical activities the internal production of heat rises steeply. The relevant considerations are set out in Figure 182.

The more active people are, the more heat they generate. If they are to remain comfortable, the air temperature within the

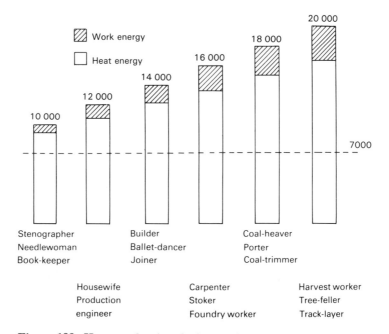

Figure 182. Heat production during various occupations.
The height of the columns, and the adjacent numbers, indicate the overall energy consumption in kJ per 24 hours; the shaded portion of each column indicates the work energy and the white portion the heat production.

Table 44. Recommended room temperatures for various activities.

Type of work	Room temperature (°C)
Sedentary mental	21
Sedentary light manual	19
Standing light manual	18
Standing heavy manual	17
Severe work	15–16

room must be lowered, so that it is easier to get rid of the
surplus heat. At a relative humidity of 50%, various kinds of
activity require the room temperatures listed in Table 44.

20.6. Heat in industry

Effects of heat

As the temperature rises above the optimum for comfort,
problems arise: first of a subjective nature, and subsequently
physical problems which impair workers' efficiency. Some of
these problems and their symptoms, in the range between a
comfortable temperature and the highest tolerable limit, are
listed in Table 45.

Table 45. Effects of deviations from a comfortable working temperature.

20°C	1.	Comfortable temperature	Maximum efficiency
	2.	Discomfort; increased irritability; loss of concentration; loss of efficiency in mental tasks	Mental problems
	3.	Increase of errors; loss of efficiency in skilled tasks; more accidents	Psycho-physiological problems
	4.	Loss of performance of heavy work; disturbed water- and salt-balances; heavy stresses on heart and circulation; intense fatigue and threat of exhaustion	Physiological problems
35−40°C	5.	Limit of tolerance of high temperature	

*Importance of
sweating*

We have seen above (Figure 176) that the range of temperatures between a comfortable level and the upper limit of
tolerance, some 10−15°C, is a zone in which heat regulation is
achieved by the evaporation of sweat from the skin. As the
external temperature rises, the body can lose less and less heat
by convection and by radiation (because of the lower temperature gradient), so that *sweating becomes the only way in
which excess heat can be lost.* In fact a point is soon reached at
which convection and radiation convey heat *into* the body, and
this heat, together with that produced internally, must be lost
by the evaporation of sweat.

When working in high temperatures, therefore, secretion

and evaporation of sweat are of paramount importance for the preservation of a correct heat balance.

Mechanisms for physiological adaptation

If the ambient temperature rises, the following physiological effects may be produced:

1. Increased fatigue, with accompanying loss of efficiency for both physical and mental tasks.
2. Rise in heart rate.
3. Rise in blood pressure.
4. Reduced activity of the digestive organs.
5. Slight increase in core temperature and sharp rise in shell temperature (temperature of the skin may rise from $32°C$ to $36-37°C$).
6. Massive increase in blood flow through the skin (from a few ml/cm^3 of skin tissue/min, to $20-30$ ml).
7. Increased production of sweat, which becomes copious if the skin temperature reaches $34°C$ or more.

The effect of these adaptive changes is clearly *to transport more heat to the skin, by means of an increased blood flow.* This increased flow is at the expense of the blood supply to the musculature (hence the reduced performance and efficiency), and to the digestive organs (which also reduce their activity). Since thermal regulation is now the overriding problem, the other systems must take second place; the muscles work less effectively and the stomach refuses food (nausea).

Similarly, the heart and circulatory system adapt themselves. The rise in blood pressure, coupled with the dilation of the blood vessels of the skin (and the simultaneous constriction of the blood vessels to the internal organs) the increased pumping action of the heart, all contribute to the increased flow of blood and transport of heat to the skin. If the skin temperature should reach $34°C$, reflex action of the heat centre produces a copious flow of sweat, which is poured out from about 2½ million sweat glands in the human skin.

Effects of overheating

If these control measures are insufficient, then the core temperature of the body begins to rise, leading to a heat accumulation which can soon be fatal. Clinical surveys have shown that during military exercises, core temperatures of around $39°C$ have resulted in heat stroke, followed by death, although Robinson and Gerking [215] reported that with rectal temperatures of $39-40°C$ heat collapse was not necessarily fatal.

Heat stroke

As the body heat rises, the first alarming symptoms include a general feeling of listlessness, loss of performance in spite of every effort, bright red skin and an increased heart rate with a feeble pulse. These are followed by severe headache, giddiness, shortness of breath, perhaps vomiting, and muscular cramps as a result of loss of salt. The final stage is unconsciousness, which may end in death within 24 hours, in spite of every medical attention, including rapid cooling of the body. Death from sunstroke is really a special case of heat stroke, in which direct heat from the sun on the head may be the decisive factor.

Liability to heat stroke is an individual matter, and varies greatly from one person to another. The risk is much greater for a fat person than for a slim one, and may be increased to six times normal if a person is 25 kg overweight. Other factors involved include capacity for heat adaptation, age, intake of food and particularly the amount of physical exercise taken. The work of Wyndham and colleagues [279] may be mentioned here. They recorded deaths from heat stroke among South African miners working in temperatures above 30°C, with 100% relative humidity; at 34·5°C the mortality was one in a thousand. Similar results were recorded for non-fatal heat collapse. Even with lower relative humidities, heat stroke sometimes occurs among the general population if the temperature outdoors rises above 35°C.

Fevers and sport   It is interesting to note that the core temperature may rise very high (41°C) during a fever, and up to 39·5°C in strenuous sporting activities, without provoking collapse through heat stroke. Under these conditions the heat centre is able to cope with the rise in core temperature, preserving the vital functions during a fever, and maintaining high performance during sport. Heat stroke is a consequence of excessive heat from outside the body, where it is passively accumulated. The brain seems to have some sort of protective mechanism against active heat accumulation generated within the body itself, but this does not operate against heat from outside.

Heat tolerance   The important question in any problem of working in excessive heat is the tolerable heat load, or heat tolerance.

From a physiological point of view, most authors agree that *rectal temperature should not exceed an upper limit of 38°C* [61, 70], and several authors have compiled indices of climatic factors which would ensure that this physiological limit is not exceeded; examples are Belding and Hatch [12], and more recently Dukes-Dobos [61]. Since these indices of heat load are very complicated, taking into account air temperature, humidity and radiant heat, we may adopt a simpler system, after Wenzel [272]. These limits of heat tolerance are shown as shaded bands in Figure 183.

Practical limits   These limits are valid for unclothed young men, who are highly efficient and well accustomed to heat. Hence they cannot be applied directly to conditions in industry! There have been no detailed investigations under working conditions, but it can be

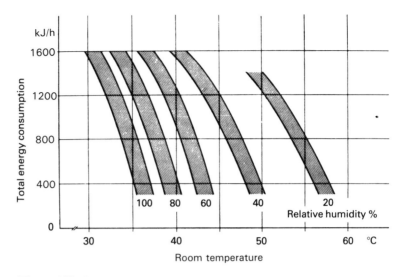

Figure 183. Limiting values for heat load in relation to physical effort (energy consumption), the relative humidity of the air and the air temperature.
Air temperature approximately the same as radiant temperature; air movements between 0·1 and 0·9 m/s. Working time 3—6 hours. Modified from Wenzel [272].

assumed that the curved bands in Figure 183 should be moved at least 5—10°C to the left to make them applicable to working conditions in industry, and even further to the left for older workers, or for jobs which require the use of protective clothing or breathing apparatus. Table 46 sets out estimated figures for acceptable temperature ranges for work in the daytime, which are applicable to capable, healthy men, wearing clothing appropriate to the job.

If the heat load is greater than those quoted in Table 46, and cannot be significantly reduced by technical means, then *the working time in the heat must be shortened.* Suggestions are given in Table 47.

Table 46. Proposed temperature limits for acceptable heat loads during daytime work.

Overall consumption of energy (KJ/h)	Examples	Upper limit of temperature (°C)	
		Effective temperature	Temp. with 50% RH
1600	Heavy work; walking, with 30 kg load	26—28	30·5—33
1000	Moderately heavy work; walking at 4 km/h	29—31	34—37
400	Light sedentary work	33—35	40—44

Fitting the task to the Man

Table 47. Permissible working times in hot, humid conditions, at heavy work of 1900 kJ/h.
After McConnell and Yaglou [182].

Wet- bulb temperature (°C)[a]	Permissible working time (min)
30	140
32	90
34	65
36	50
38	39
40	30
42	22

[a] *The wet-bulb temperature is the reading of a thermometer with its bulb covered with a wick dipping into water. Its reading is lower than air temperature because of increased evaporation.*

Working under radiant heat

The limiting values quoted in Figure 183 and Tables 46 and 47 apply to conditions where the surrounding surfaces are at approximately the same temperature as the air. They cannot be applied without modification to workplaces that are exposed to great radiant heat, such as, for example, in front of furnaces. Radiant heat is often measured by means of a *globe thermometer*. This usually consists of a hollow sphere of copper, about 150 mm in diameter, into which a normal mercury thermometer is inserted. The outer surface of the copper sphere is painted black, to absorb the radiation and to convert it into heat. *The globe thermometer takes up an average value between the temperature of the air and that of surrounding surfaces.*

Various methods have been devised to integrate the various temperature factors into one. One of these is to combine the wet-bulb and globe thermometer readings into the WBGT (wet-bulb−globe temperature). WBGT=0·7 wet-bulb reading+0·3 globe thermometer reading. Using this formula, the following values of WBGT for continuous work were proposed in the USA:

$$2090 \text{ kJ/h} \quad \text{WBGT of } 25°C$$
$$1460 \text{ kJ/h} \quad \text{WBGT of } 27°C$$
$$840 \text{ kJ/h} \quad \text{WBGT of } 30°C$$

Physiological limits

In practice it is usually very difficult to express the heat load and the degree of physical work exactly in kJ, so physiological means have to be sought to assess these. As we have already seen, heart rate, core temperature (rectal temperature) and perspiration can all be used for this purpose. Upper limits of these parameters for working in heat for an entire working day are:

1. Heart rate (daily average):   100—110 beats/min
2. Rectal temperature: 38°C
3. Evaporation of sweat: 0·5 l/h.

Acclimatization to heat

Experience shows that workers take time to become accustomed to working under very hot conditions, and only after several weeks is their performance equal to that of workers who are already 'heat-adapted'. This is a genuine acclimatization of the body to heat, which proceeds by the following steps:

1. The body gradually increases its perspiration, losing more and more heat in the process. A worker who is heat-adapted can lose 2 l of sweat per hour, and up to 6 l per working day.
2. As part of the acclimatization process the sweat becomes more dilute, with a lower concentration of salts. The sweat glands 'learn' to conserve salts, so that larger amounts of sweat can be produced without creating a salt deficit in the body. Such a deficit would lead to muscular cramps and eventually to exhaustion, and possibly death.
3. During acclimatization there is a loss of weight, which helps heat loss by reducing the amount of insulating fat, and also reduces energy consumption.
4. As acclimatization proceeds, the worker drinks more fluids, to compensate for the greater amount of water lost by sweating.
5. The blood system and heart also adapt themselves to provide for the improved performance after acclimatization.

When a worker has become acclimatized to heat he feels thirsty whenever his body needs more liquid, and so he tends to drink small amounts frequently. The question whether he should take additional salt is debatable. In the USA good results have been obtained from giving either salt tablets or salt-rich foods (meat broth, etc) both to factory workers and to troops exposed to excessive heat. In Europe heat stresses are not usually as great as they are in the USA, so that up to the present the provision of additional salt has not been necessary.

Recommendations

The following guidelines may be recommended for workers in very hot conditions:

1. A worker should be acclimatized to heat by stages. He or she should begin by spending only 50% of the working time in the heat and increase this by 10% each day. The same procedure should follow when a worker returns from illness or holiday.
2. The greater the heat load and the greater the physical effort performed under heat stress, the longer and more

frequent should be the pauses (cooling periods). If the limit of heat tolerance is being exceeded, the working day must be shortened.

3. The worker should drink small amounts of fluid at frequent intervals; never more than 0·25 l at a time, and a cupful every 10−15 minutes is recommended.

4. As drinks we recommend lightly sweetened tea or coffee, varied with an occasional drink of soup. If large quantities of fluid are needed it is best to drink plain water, with an occasional drink of tea or coffee. The beverages should be lukewarm or warm, so that they are more quickly absorbed by the digestive system.

5. Iced drinks, fruit juices and alcoholic drinks are not to be recommended. Milk drinks, too, are unsuitable for working in hot conditions, because they put more stress on the digestive organs.

6. The drinks should be available close to the worker, so that he or she can drink whenever he or she feels the need.

7. Where the radiant heat is excessive (e.g., blast furnaces) the worker must be protected by special goggles, screens and protective clothing against the risk of burning the eyes and hands.

In addition to these personal precautions, every attempt must be made to reduce the impact of the heat on the worker. Some possible measures include improving ventilation, both natural and forced, perhaps artificial drying of the air, and screens to protect against radiant heat.

20.7. Air pollution and ventilation

Deterioration of air

If a room has people in it, the air undergoes deterioration in various ways, changing its character due to:

1. *Release of odours.*
2. *Formation of water vapour.*
3. *Release of heat.*
4. *Production of carbon dioxide.*
5. *Air pollution*, either entering from outside, or generated by activities within the room.

The first four of these arise mainly from the human body itself. The last, air pollution, depends on the situation of the building, what activities are carried on indoors and whether smoking takes place or not. Among the air changes that are of human origin, the odours given off by the skin are most important, since in very low concentrations they give rise to feelings of unpleasantness, disgust, distaste and revulsion. These odours are a mixture of organic gases and vapours, which are not toxic in the concentrations usually encountered,

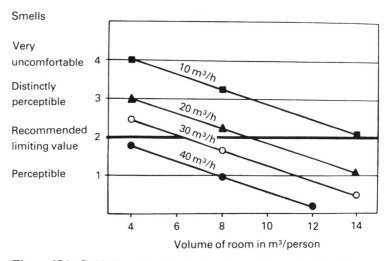

Figure 184. Guidelines for the fresh air requirements of sedentary workers in relation to the air space available to each person. *After Yaglou* et al. [280].

Table 48. Recommended air and fresh air volumes per person.

Volume per person	Fresh air per person (m³/h)	
(m³)	Minimal	Desirable
5	35	50
10	20	40
15	10	30

but which are most undesirable on account of their subjective annoyance. When the air pollution in a room is primarily human in origin, it is the personal odours that are most offensive, far more than the changes in carbon dioxide or water vapour. A general guideline for the quantity of fresh air needed in a room with people in it is *30 m³ of fresh air per person per hour.*

But this is only a very rough figure. Figure 184 summarizes the results obtained by Yaglou and his colleagues [281] in a long series of experiments.

From these results we can make the recommendations of Table 48.

Environmental tobacco smoke

In recent years increasing importance has been attributed to environmental tobacco smoke. This is mainly due to its acknowledged long-term health risks, but on a day-to-day basis it also often causes annoyance and even eye and throat irritations in workrooms as well as in vehicles and restaurants.

Annette Weber [269] studied the effects of cigarette smoke on non-smoking subjects, so-called *passive smoking*. A field study on 472 employees in 44 workrooms revealed that 64% of the non-smokers were "sometimes or often" disturbed by the environmental tobacco smoke, whereas 36% reported eye irritations at work. Approximately one-third of all employees qualified the air at work with regard to smoke as bad. The measured carbon monoxide (CO) concentrations due to tobacco smoke varied between 0 and 6·5 ppm with a mean value of 1·1 ppm.

In a second step, several experimental studies were carried out in a climatic chamber in which cigarettes were smoked by a smoking machine. The degree of air pollution due to tobacco smoke was evaluated by measuring the concentrations of carbon monoxide (CO) and several other pollutants such as nitric oxide, formaldehyde, acrolein, particles and nicotine. The degree of acute irritating and annoying effects was determined simultaneously on exposed subjects by means of questionnaires and measurements of the eye-blink rate.

The results revealed that the irritating effects were most pronounced on the eyes, followed by the effects on the nose and on the throat. Figures 185 and 186 illustrate the results obtained for subjective eye irritations and eye-blink rate of

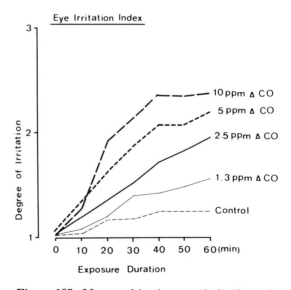

Figure 185. Mean subjective eye irritations due to environmental tobacco smoke, related to smoke concentration and duration of exposure.
$\Delta CO = CO$ *level during smoke production minus background level before smoke production. 32–43 subjects. 0 min = measurement before smoke production. Period 0–5 min = increasing smoke concentration. Period 6–60 min = constant smoke concentration. Degree of irritation:* 1 = no *irritation;* 2 = little *irritation;* 3 = moderate *irritation.*

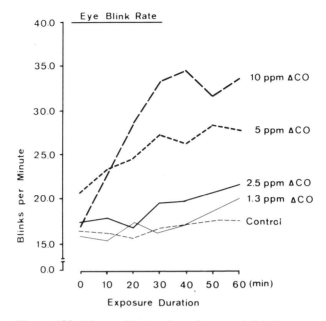

Figure 186. Mean effects of environmental tobacco smoke on eye blink rate.

ΔCO = CO level during smoke production minus background level before smoke production. 32−43 subjects. 0 min = measurement before smoke production. Period 0−5 min = increasing smoke concentration. Period 6−60 = constant smoke concentration.

people being exposed to different smoke concentrations which were kept constant for nearly one hour.

It is evident that the mean eye irritation as well as the blink rate increase with higher smoke concentration and with the duration of exposure. The mean incidences of subjects with eye irritations are reported in Figure 187. The results clearly reveal that there is a marked increase in strong eye irritations between the smoke levels corresponding to 1·3 and 2·5 ppm ΔCO. The author concludes that a possible limit to protect healthy people in their everyday environment against adverse effects of environmental tobacco smoke should lie in this range, i.e., between 1·5 and 2·0 ppm ΔCO. Indeed, the Figure shows a marked increase from about 3 to over 10% of the subjects with strong eye irritations. The upper concentration limit of 2·0 ppm ΔCO, for instance, is already reached when two cigarettes are smoked per hour in a room of 80 m³ with a single air change.

Hence counter-measures to protect passive smokers are desirable when the ΔCO level reaches 1·5 ppm and are necessary when it hits 2·0 ppm. The lower limit should be applied to workplaces where passive smokers can hardly escape the exposure, and the upper limit to restaurants and other places, where people usually go voluntarily and for a shorter lapse of time.

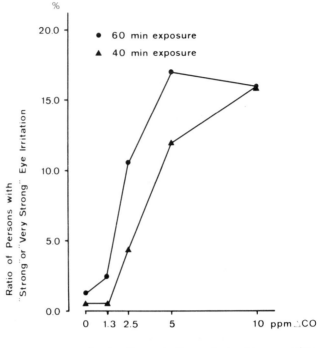

Figure 187. **Percentage of persons with strong or very strong eye irritation reactions related to the degree and duration of exposure.**

Fresh air supply in smoking rooms

Calculations showed that *a fresh air supply of 33 m³/h and per smoked cigarette is necessary to keep the ΔCO concentration below the proposed upper limit of 2·0 ppm; for the lower limit, 50 m³/h of fresh air per smoked cigarette are required.* Depending on the number of people present in a room, a fresh air supply of 25—45 m³/h/person is necessary in order not to exceed the upper limit. In other words: the ventilation has to be 2—4 times higher than in a room where nobody smokes (in which only 12—15 m³/h/person are required). This increased ventilation as a measure to protect passive smokers is of course not recommendable from the energetical and economical point of view. Therefore, whenever possible, organizational measures, such as separation into smoking and non-smoking rooms or a prohibition of smoking, rather than an increase of fresh air supply, should be taken into consideration.

Organizational measures

Natural or forced ventilation

After internal pollution, the situation of the building and the window area are decisive factors in deciding whether it is necessary to have any form of forced ventilation, or air conditioning. If it is impossible to open the windows in summer, either because of traffic noise or because of the pollution of the urban air, then artificial ventilation of some kind is essential.

Windows and climate indoors

Modern buildings tend to have lower ceilings, and more window area then buildings erected 50 or more years ago. These changes have important effects on the climate indoors. Lower ceilings make fresh air even more necessary. Huge window areas act as cooling surfaces in winter, and from spring to autumn they allow a great deal of heat to penetrate into the room. This becomes particularly obvious if K values (thermal conductivity) are used as a measure of the heat balance of a building. The windows of older buildings occupied 15−30% of the external surface, whereas they are often more than 50% of the surface of modern buildings. If we assume a K value of a window as being $3 \cdot 5$, and that of a wall as $0 \cdot 8$, then if the windows occupy 50% of the outer walls, it can be calculated that 82% of the total heat loss occurs from the windows, and only 18% from the walls and roof.

This means that in such a building four times as much heat is lost through the windows as through the walls and roof. Improving the thermal insulation of the window is four times as effective as insulating the roof and walls, and is greatly to be recommended.

Modern buildings admit more light and bring the occupants into closer contact with the outside world, with the sky, and with nature; at the same time they raise serious problems of controlling the climate indoors and lead to a greater dissipation of energy.

21. Daylight, colours and music for a pleasant work environment

21.1. Daylight

The physiological demands of daylight are essentially similar to those of artificial light, but there are some special problems, which will now be briefly discussed.

Besides providing illumination, natural daylight penetrating into a room establishes contact with the world outside, giving a view of the surroundings and indicating the time of day and the state of the weather.

It is psychologically and physiologically desirable to have as much light, as evenly distributed, as possible. The higher the daylight level, the less need there is for artificial lighting, especially in winter. On a cloudy day in December, the daylight intensity indoors will reach 500 lx only for the four hours from 10·00 to 14·00 hours.

The daylight illumination levels depend mainly on the position and type of windows; the height of windows is decisive for the penetration of light into the depth of the room.

One reservation must be made. Big, high windows certainly help to distribute daylight within a room, but they have the drawback that they admit a great deal of summer heat, especially if they face south or south-west. Furthermore, in winter they act as cold surfaces, with an adverse effect on the temperature of the room. The sizes of windows must not be decided solely in relation to daylight; there must be a 'balance sheet' of all the pros and cons.

Recommendations

The following are nine rules of thumb relating to daylight indoors.

1. *High windows are more effective than broad ones*, since the light penetrates further into the room. The lintel should not be deeper than 300 mm.
2. *Window sills should be at table height.* If the window extends below the table-top it is cold in winter and may cause glare.

3. *The distance from window to workplace should not be more than twice the height of the window.*
4. *For workrooms the window area should be about one-fifth of the floor area.* This is only a general rule which is very flexible according to circumstances.
5. *It is important that the glass should transmit plenty of light.* Clear glass has a transparency of more than 90%, whereas frosted glass, glass bricks, or special heat-insulating glass may have transparencies from 70% down to only 30%.
6. *Effective protections against the glare of direct sunlight, and against radiant heat,* are important in securing good visibility and comfort indoors. *The most efficient method is an adjustable external sunshade,* either venetian blinds or shutters. Venetian blinds inside the window, or between the panes of double-glazing, are a mistake, because they afford no protection against radiant heat.

   Insulating window panes are not adjustable; they reduce the amount of light transmitted, which is undesirable in winter, while in summer they trap heat inside the room (greenhouse effect). Balconies, overhanging eaves and other projections also create problems, by cutting off some of the light in dull weather. In our latitudes they should be restricted to south-facing walls.
7. *Each window should receive direct light from the sky, and it is desirable that a portion of sky should be visible from every workplace.*
8. The nearest building should be at least twice as far away as its own height.
9. Pale colours should be used, both in the room itself, and in any courtyard outside, so as to reflect as much of the incident daylight as possible.

Glass houses and windowless factories

The trend in modern architecture has been to increase the window area of new buildings, sometimes ending up with a glass house. As we have already said, glass walls, or very large window areas, *radiate heat away in winter and admit excessive heat in summer.* On the other hand they admit more daylight and allow occupants to see more of the outside world. The problems of regulating the room climate are both difficult and expensive to solve. However, it must be said that recent developments in glass technology have reduced these problems considerably.

Equally problematical are factory buildings that are not provided with windows at all, on grounds either of economy or increased production. The arguments in favour of such buildings are that they can be given a uniform internal climate (through air-conditioning), and that the internal lighting, being entirely artificial, can be controlled by modern techniques. The more complete insulation of the walls reduces the cost of air-conditioning. Against these methods it is argued that the absence of windows makes the workers feel 'imprisoned' and 'cut off', and that people need some contact with the outside world when they are at work. Since the validity of these various arguments is not yet proved it would seem sensible to proceed slowly with building practices that are so unnatural, and so psychologically questionable. The requirement of some authorities that at least some windows should be provided seems a sensible interim provision.

Skylights and fanlights

Skylights in the roof and fanlights in the walls or door are often useful accessories in single storey buildings, in lofts and attics, and in any other rooms where insufficient daylight penetrates. The following are the most important types, from the standpoint of industrial physiology:

1. *Pitched roofs.* If the axis of the skylight is no further away from the workplace than the height up to the ridge, the lighting is bright and even. It is particularly good if the workplaces are so arranged that each lies between two skylights at equal angles, and so is lit from both sides.
2. *Skylights in roof ridge.* This gives glass above the longitudinal axis of the building. It is particularly suitable for high workshops, where the operatives in the middle need the best light.
3. *Mansard roofs.* Here the skylights are placed in the sloping part of the mansard roof, above the windows, giving a relatively high level of daylight all along the outer walls. However, they increase the risk of glare and do not give much extra light to the people in the middle of the room.
4. *Zig-zag shed roofs.* The whole working area has clear glass above it, in panels which usually face north and slope upwards at an angle of $60°$. The reverse face of each zig-zag is opaque, faces south, and slopes at an angle of $30°$. This arrangement gives the highest and most uniform values of daylight quotient, and is especially recommended for large workshops.

21.2. Colour in the workplace

Physical principles

The accepted colours of the spectrum cover the following bands of wavelength, in nanometres ($nm = 10^{-9} m = $ one millionth of a mm).

Violet:	380−436
Blue:	436−495
Green:	495−566
Yellow:	566−589
Orange:	589−627
Red:	627−780

Electromagnetic waves longer than 780 nm belong to the infra-red (radiant heat); shorter than 380 nm, they are ultra-violet rays, which are of critical importance for the synthesis of vitamin D from its precursor ergosterol and for normal organic growth.

The colours that we see arise because the molecular structure of the surface of objects reflects only part of the light that falls on it. This is what we perceive. For example a green-painted machine absorbs all the incident light except the green. The colour receptors in the retina are the cones, which are capable of distinguishing more than 100 000 shades of colour.

Reflected colour

When deciding upon the colours at and around a workplace, it is necessary to consider reflectivity. Table 49 gives a few examples.

Colours in and around workplaces have the following functions:

1. *Orderliness, and as an aid to identification.*
2. *To indicate safety devices.*
3. *Contrasts to make work easier.*
4. *A psychological effect on the operator.*

Table 49. Reflectance as a percentage of the incident light.

Colour or material	Reflectance (%)
White	100
Aluminium; white paper	80−85
Ivory; deep lemon yellow	70−75
Deep yellow; light ochre; light green; pastel blue; pale pink; cream	60−65
Lime green; pale grey; pink; deep orange; bluegrey	50−55
Powdered chalk; pale wood; sky blue	40−45
Pale oakwood; dry cement	30−35
Deep red; grass green; wood; pale leaf green; olive green; brown	20−25
Dark blue; purple red; reddish brown; slate grey; dark brown	10−15
Black	0

Orderliness

Certain rooms, or floors, or sections of the factory can be given a colour code, which helps to keep the whole works on an orderly plan. In very big, scattered industrial plants, or those which are under multiple direction, different colours can be used to introduce a certain degree of orderliness and to facilitate supply and servicing.

Safety colours

If the same colour is always used to indicate a particular danger, the correct reaction to it becomes automatic, so this practice is now followed in most countries. More details will be found in DIN Standards 4844 and 5381 [56] or in Hettinger *et al.* [114].

These are some of the common colour codes:

1. *Red* is the '*danger colour*': Halt, Stop, Prohibited. Red is also the warning colour for Fire; on extinguishers and other equipment.
2. *Yellow*, usually in contrast with Black, means *Danger of collision*, *Look out*, *Risk of tripping*. Yellow and Black are much used as warning colours in transport.
3. *Green* means *rescue services*, *safety exit*, etc. It is used to indicate all forms of rescue equipment and first aid.
4. *Blue* is not actually a safety colour, but is used for *giving directions*, advice, signs, etc.

Colour contrasts of large areas

When deciding on colour contrasts, large areas such as walls and furniture must be considered separately from small areas such as splashes of colour intended to attract attention on knobs, handles, levers, etc.

The colours of large areas should be chosen so that they have similar reflectivities (see Table 49), so as to have colour contrast without differences in brightness. The avoidance of large areas of contrasting brightness, close together, is an important factor in ensuring good visual acuity. Large areas and big objects should never be painted in pure colours, nor with fluorescent paint, since these cause local overloading of the retina, and lead to the production of after-images. So, walls, partitions, table tops, etc. should be painted with unsaturated colours, in a matt finish.

It is easier to lay out working materials and to select the one required, if they are coloured differently from their immediate surroundings, and this should be borne in mind when the workplace is being designed. At the same time contrasts in brightness and intensity of colour should be avoided. For example, if the working materials are made of leather, wood, or similar materials of an ochre yellow or brown colour, a suitable colour for the background would be matt green, nile green, or matt bluish. Greyish-blue materials such as steel and other metals show up well against a background of dark ivory or light beige. The area surrounding the machine, or the work table, might be painted in cool, neutral colours, from yellow green to pastel blue.

Eye-catching colours

Eye-catching colours are the little spots of strongly contrasting colours that are used to attract attention, to 'catch the eye'. Colour is used in this way in nature: a red strawberry among green foliage; brilliant flowers which attract insects and other creatures by their colour contrast. This effect is as important to animals as it is to plants. On the other hand nature also uses colours for concealment. Defenceless wild animals are often neutral in colour, and merge into their background, so that they are almost invisible.

It is a good idea to provide a few eye-catchers at a workplace, marking such things as the more important handles, levers, control-wheels, knobs, and so on. If the eye-catchers are small, not more than a few square centimetres in area, they should contrast strongly, not only in colour, but also in brightness. Eye-catchers make the controls easier to find, reducing the time taken to search for them, and hence the diversion of attention from the work itself.

The human eye sees the greatest contrast between yellow and black, because the brain adds together the effect of colour and intensity. So yellow—black contrast is recommended for indicator instruments as well as for switch panels.

The greatest danger in colour planning, and especially in the planning of eye-catchers, is excess. If there are too many eye-catchers, in too many different colours, then the whole workplace becomes restless and distracting. Colour does not mean bunting! *The most important physiological requirement in the use of colour is restraint, with three, or at most five eye-catchers to each work-place.* This applies, too, to colour in schoolrooms, restaurants, homes — everywhere, in fact, where people either work or relax. Less restraint is appropriate in shop-windows, display cases and exhibitions, where the customer is meant to be stimulated by eye-catchers.

Psychological effects

By the 'psychological effects' of colour we mean optical illusions and other phenomena that are triggered off by colour.

In part these are caused by subconscious associations with previous sights or experiences, and partly by hereditary factors. They influence mental affectivity, and thereby all of a person's behaviour. ('Events' in art are phenomena of this kind. Modern abstract painting strives to produce such effects by colour and form alone, which, to the 'initiated', are at least as stimulating emotionally as representational pictures.)

Psychological effects can be induced, too, by colours in a room, arousing strong feelings of like or dislike. Since, however, rooms have to serve particular functions, their colours do not only have aesthetic consequences: their physiological and psychological effects must also be taken into account. Yet there is always a good deal of latitude for aesthetic considerations.

Particular effects

Particular colours have special psychological effects, which are all more or less similar in character, although with great individual variation. The most important particular effects concern distance, temperatures and the effects on the mental state. Table 50 summarizes these illusory effects of individual colours.

Table 50. Psychological effects of colour.

Colour	Distance effect	Temperature effect	Mental effect
Blue	Further away	Cold	Restful
Green	Further away	Cold to neutral	Very restful
Red	Closer	Warm	Very stimulating, not restful
Orange	Much closer	Very warm	Exciting
Yellow	Closer	Very warm	Exciting
Brown	Much closer, claustrophobic	Neutral	Restful
Violet	Much closer	Cold	Aggressive; unrestful, tiring

Broadly speaking, all dark colours are oppressive and tiring; they absorb the light and are difficult to keep clean. All light colours are bright, friendly and cheerful; they scatter more light, brighten up the room, and encourage greater cleanliness.

Colours in a room

Before starting to plan the colour for a room, there must be a careful consideration of its functions, and who is going to use it. After that it will be possible to plan its colours in relation to psychological and physiological factors.

The principles mentioned above must be adhered to. Consideration must be given to the work to be carried on in it, whether this is likely to be monotonous, or whether it will make heavy demands on the concentration. If the work is monotonous, it is advisable to include a few areas of exciting colour, but not big areas such as the main walls or ceilings; merely a few items such as a pillar or column, a door, or a partition wall.

If the workroom is very large, it can be divided up by the use of different colours, thereby making it less anonymous.

If the work going on in the room demands close concentration, the colours should be chosen carefully so as to avoid unnecessary distractions and unrestful items. In this case walls, ceilings and other structural elements should as far as possible be painted in light colours that do not attract attention.

Walls and ceilings painted yellow, red or blue may be very attractive at first glance, but as time goes on they become a strain on the eyes. Hence such rooms often become unpleasant after a while.

More intense colours can safely be used in rooms that are mainly used briefly, e.g., entrance halls, corridors, lavatories, storerooms, etc. Here strong colours may help to brighten up the room, and make it structurally more pleasing. Colour plans for schools, hospitals, and administrative blocks should be based on similar principles.

21.3. Music and work

Throughout the ages, music has been used more and more to lighten human labours, and many working songs exist; some of the best-known being those of spinning women, soldiers' marching songs, and the famous Song of the Volga Boatmen. All songs of this kind are melodious, with a well-marked rhythm, and their effect is to rouse the singers and urge them on to greater effort.

Physiological effects

An acoustic stimulus passes along the auditory nerve to the activating system and brings the entire conscious sphere of the cerebral cortex into a state of advanced readiness for action. Hence noise can have a stimulating effect, especially in boring situations. Music that is strongly rhythmical, with marked variations in loudness, affects the brain in similar ways, bringing the whole organism to 'action stations'. Noise also has a distracting effect, however, so that activities that call for thought and alertness are disturbed. We might expect the same to be true of stirring, rhythmical music. In fact we do find that music is particularly welcome as a background to dull, repetitive work, but its effect on intellectual work is debatable.

The extent of the distraction and disturbance provoked by music depends a good deal on its nature. Up to a point the distraction can be minimized by choosing suitable music.

Studies in industry

As part of the campaign to improve working conditions, for more than 30 years music has been used, here and there, to relieve the boredom of certain jobs. Thus English research showed that the introduction of music into a factory making ready-made clothing improved the output by female workers. From their experience, the investigators recommended that the music should be restricted to a period in the mornings, between 10·00 and 11·15 am. Kerr [142] questioned 666 workers in a US factory, whose jobs included spool-winding, operating presses and assembling radio tubes, about how they would like music distributed throughout the day. The overwhelming majority wanted continuous music all day long; if it had to be restricted, then most of them preferred a series of between 10 and 16 periods, equally distributed throughout the day. Mid-morning and mid-afternoon were periods when music was particularly welcome. It appeared that the younger employees and the women liked music more than the others.

Similar research was carried out in French factories, which by and large confirmed these results. In Switzerland music was introduced in textile mills, shoe factories and the like, with similar results.

Background music

Originally, music at work was rhythmical, with a clear melody. The workers listened to it consciously, and sometimes hummed the tune. More recently, starting in the USA, administrative offices, business houses, salerooms, railway stations, waiting rooms, restaurants and even residential rooms have been provided with a different kind of music, persistent, but very quiet, unobtrusive, hardly impinging on the consciousness. This is 'background music' or 'muzak', which is supposed to surround one with a pleasant 'envelope' of agreeable sound, which should have the advantage of not being distracting, and therefore being suitable for work which demands concentration, such as designing or planning.

Recommenda-tions

As far as present knowledge goes, the question of whether music at work is a good thing or not may be answered as follows.

Music at work helps to create a pleasant atmosphere, which stimulates the worker. This is particularly so if the work is boring, or repetitive, or makes few demands on thought or alertness. Music is less helpful in big, noisy workshops, or in jobs where mental alertness is essential.

When the music chosen is not quiet, background music, but assertive music that needs to be listened to, it should be played only for part of the working day. A short period of rousing music should begin the day, which should end with more festive tunes, while the rest of the day should have four periods of 30 minutes each of light music.

The tempo should be neither too slow and soporific, nor too fast causing irritation and haste.

References

1. Aanonsen, A.: *Shiftwork and Health*. Norwegian Monographs on Medical Science, Universitets forlaget, Oslo (1964).
2. Abramson, N.: *Information Theory and Coding*. McGraw-Hill, New York (1963).
3. Akerblom, B.: *Standing and Sitting Posture*. Nordiska Bokhandeln Stockholm (1948).
4. Allensbacher Berichte: *Wer will die Viertagewoche?* Institut für Demoskopie Allensbach, D-7753 Allensbach am Bodensee, Nr 16 (1971).
5. Andersson, B. J. G. and Ortengren, R.: Lumbar disc pressure and myoelectric back muscle activity during sitting. 1. Studies on an office chair. *Scandinavian Journal of Rehabilitation Medicine* 3, 115—121 (1974). The same author with colleagues also in *Scandinavian Journal of Rehabilitation Medicine* 3, 104—114 (1974), 3, 122—127 (1974), 3, 128-135 (1974).
6. Andlauer, P. and Fourré, L.: Aspects ergonomiques du travail en équipes alternantes. Rapport du Centre d'études de Physiologie appliquée au travail, Strasbourg (1962) and in: *La Revue Française du Travail*, 35—50 (1965).
7. Andlauer, P., Carpentier, J. and Cazamian, P.: *Ergonomie du Travail de Nuit et des Horaires Alternants*. Editions Cujas, Paris (1977).
8. Barnes, R. M.: An Investigation of Some Hand Motions Used in Factory Work. University of Iowa, Iowa City, *Studies in Engineering*, Bulletin 6 (1936).
9. Barnes, R. M.: *Motion and Time Study*. 3rd edition, John Wiley, New York (1949).
10. Barnes, R. M., Hardaway, H. and Podalsky, O.: Which pedal is best? *Factory Management and Maintenance*, 100, 98—99 (1942).
11. Baschera, P. and Grandjean, E.: Effects of repetitive tasks with different degrees of complexity on CFF and subjective state. *Ergonomics*, 22, 377—385 (1979).
12. Belding, H. S. and Hatch, T. F.: Index for evaluating heat stress in terms of resulting physiological strains. *Heating, Piping and Air Conditioning*, 27, 129—135 (1955).
13. Bell Telephone Laboratories, Inc.: *Video Display Terminals — Preliminary Guidelines for Selection, Installation and Use*.
14. Bendix T. and Hagberg M.: Trunk posture and load on the trapezius muscle whilst sitting at sloping desks. *Ergonomics*, 27, 873—882 (1984).

15. Benz, C., Grob, R. and Haubner, P.: *Designing VDU Workplaces.* Verlag TÜV Rheinland Köln (1983). Deutsche Ausgabe: Gestaltung von Bildschirm-Arbeitsplätzen.

16. Bhatnager, V., Drury, C. G. and Schiro, S. G.: Posture, postural discomfort and performance. *Human Factors*, **27**, 189-199 (1985).

17. Bills, A. G.: Blocking: a new principle of mental fatigue. *American Journal of Psychology*, **43**, 230-239 (1931).

18. Bjerner, B., Holm, A. and Swensson, A.: Diurnal variation in mental performance. *British Journal of Industrial Medicine*, **12**, 103−110 (1955).

19. Blackwell, H. R. and Blackwell, O. M.: The effect of illumination quantity upon the performance of different visual tasks. *Illuminating Engineering*, **63**, 143−152 (1968).

20. Blum, M. L. and Naylor, J. C.: *Industrial Psychology.* Harper and Row, New York (1968).

21. Bonvallet, M., Dell, P. and Hiebel, G.: Tonus sympathique et activité électrique corticale. *Journal of Electroencephalography and Clinical Neurophysiology*, **6**, 119−125 (1954).

22. Bouisset, S.: Postures et mouvements. In *Physiologie du Travail*, Tome 1, ed. by J. Scherrer. Masson, Paris (1967).

23. Bouisset, S. and Monod, H.: Etude d'un travail musculaire léger. I. Zone de moindre dépense energétique. *Archives Internationale du Physiologie et Biochimie*, **70**, 259−272 (1962).

24. Bouisset, S., Laville, A. and Monod, H.: Recherches physiologiques sur l'économie des mouvements. Proceedings of 2nd IEA Congress, *Ergonomics*, **7**, 61−67 (1964).

25. Bouma, H.: Visual reading processes and the quality of text displays. In *Ergonomic Aspects of Visual Display Terminals*, ed. by E. Grandjean and E. Vigliani. Taylor & Francis, London (1980).

26. Bräuninger, U., Grandjean, E., van der Heiden, G., Nishiyama, K. and Gierer, R.: Lighting characteristics of VDTs from an ergonomic point of view. In *Ergonomics and Health in Modern Offices*, ed. by E. Grandjean. Taylor & Francis, London (1984).

27. British Standards Institution: Guide to the evaluation of human exposure to whole-body vibration. No. DD 32, London (1974).

28. Broadbent, D. E.: Effects of noise on behaviour. In *Handbook of Noise Control*, ed. by C. M. Harris. McGraw-Hill, New York (1957).

29. Broadbent, B. E.: *Perception and Communication.* Pergamon Press, London (1958).

30. Broadbent, D. E.: Effect of noise on an intellectual task. *Journal of the Acoustical Society of America*, **30**, 824−827 (1958).

31. Brouha, L.: *Physiology In Industry*, 2nd edition. Pergamon Press, Oxford (1967).

32. Brown, J. S. and Slater-Hammel, A. T.: Discrete movements in the horizontal plane. *Journal of Experimental Psychology*, **39**, 84−95 (1949).

33. Brown, J. S., Knauft, E. B. and Rosenbaum, G.: The accuracy of positioning reactions as a function of their direction and extent. *American Journal of Psychology*, **61**, 167−182 (1948).

34. Brundke, M.: Langzeitmessungen der Pulsfrequenz und Möglichkeiten der Aussage über die Arbeitsbeanspruchung. In: *Pulsfrequenz und Arbeitsuntersuchungen*, Schriftenreihe Arbeitswissenschaft und Praxis, Band 28. Beuth-Vertrieb, Berlin (1973).

35. Buesen, J.: Product development of an ergonomic keyboard. Proceedings of Ergodesign 84, *Behaviour and Information Technology*, **3**, 387−390 (1984).

36. Cakir, A., Reuter, H. J., von Schmude, L. and Armbruster, A.: Anpassung von Bildschirmarbeitsplätzen an die physische und psychische Funktionsweise des Menschen, *Bundesministerium für Arbeit und Sozialordnung*, Referat Presse, Postfach, 5300 Bonn (1978).

37. Cakir, A., Franke, R. P. and Piruzram, M.: Arbeitsplätze für Phonotypistinnen. *Bundesanstalt für Arbeitsschutz*, Report No. 363, Dortmund (1983).

38. Caldwell, L. S.: The effect of the special position of a control on the strength of six linear hand movements. US Army Medical Research Laboratory, Fort Knox, Kentucky, Report Nr 411 (1959).

39. Caplan, R. D., Cobb, S., French, J. R., Harrison, R. V. and Pinneau, S. R.: Job demands and Worker Health: Main Effects and Occupational Differences. Institute for Social Research, University of Michigan, Ann Arbor (1980).

40. Chaffin, D. B.: A computerized biomechanical model development of and use in studying gross body actions. *Journal of Biomechanics*, **2**, 429−441 (1969).

41. Chaffin, D. B.: Localized muscle fatigue − definition and measurement. *Journal of Occupational Medicine*, **15**, 346−354 (1973).

42. Chaffin, D. B. and Andersson G.: *Occupational Biomechanics*. John Wiley, New York (1984).

43. Chaney, R. E.: Subjective Reaction to Wholebody Vibration. Boeing Company, *Human Factors Technical Report* D3, 64−74, Wichita, Kansas (1964).

44. Chief Inspector of Factories: *Annual Report 1981*. HMSO, London (1981).

45. Christensen, E. H.: *L'Homme au Travail*. Sécurité, Hygiène et Médecine du Travail, Series No. 4. Bureau International du Travail, Geneva (1964).

46. Clarke, H. H., Elkins, E. C., Martin, G. M. and Wakim, K. G.: Relationship between body position and the application of muscle power to movements of joints. *Archive of Physical Medicine*, **31**, 81−89 (1950).

47. Collins, J. B.: The role of a sub-harmonic in the wave-form of light from a fluorescent lamp in causing complaints of flicker. *Ophthalmologica*, **131**, 377−387 (1956).

48. Colquhoun, W. P. (ed.): *Biological Rhythms and Human Performance*. Academic Press, London (1971).

49. Conroy, R. T. W. L. and Mills, J. N.: *Human Circadian Rhythms*. Churchill, London (1970).

50. Cox T.: The nature and measurement of stress. *Ergonomics*, **28**, 1155−1163 (1985).

51. Damon, A., Stoudt, H. W. and McFarland, R. A.: *The Human

Body in Equipment Design. Harvard University Press, Cambridge, MA (1966).

52. Davies, D. R. and Tune G. S.: *Human Vigilance Performance.* Staples Press, London (1970).

53. Davis, P. R. and Stubbs, D. A.: Safe levels of manual forces for young males. *Applied Ergonomics,* **8,** 141−150 (1977).

54. Davis, P. R. and Stubbs, D. A.: A method of establishing safe handling forces in working situations. Report of the International Symposium on Safety in Manual Materials Handling. NIOSH (1977).

55. DIN Standard 4549: *Schreibtische, Büromaschinentische und Bildschirmarbeitstische.* Beuth-Verlag, Berlin (1981).

56. DIN Standards No. 4844 und No. 5381: *Sicherheitskennzeichnung, II. Sicherheitsfarben und Kennfarben.* Beuth-Vertrieb, Berlin (1977) and (1976).

57. DIN Standard No. 5035: Blatt 1 und Blatt 2: *Innenraumbeleuchtung mit künstlichem Licht.* Beuth-Vertrieb, Berlin (1972).

58. DIN Standard No. 33401: *Stellteile.* Entwurf 1974. Beuth-Vertrieb, Berlin (1974).

59. Drury, C. G. and Francher, M.: Evaluation of a forward-sloping chair. *Applied Ergonomics,* **16,** 41−47 (1985).

60. Dubois-Poulsen A.: Notions de physiologie ergonomique de l'appareil visuel. In *Physiologie du Travail,* Tome 2, pp. 114−183, ed. by J. Scherrer. Masson, Paris (1967).

61. Dukes-Dobos, F. N.: Rationale and provisions of the work practices standard for work in hot environments as recommended by NIOSH. In *Standards for Occupational Exposures to Hot Environments,* ed. by S. M. Horvath and R. C. Jensen. US Dept. Health, Education and Welfare, NIOSH, No. 76−100, Cincinnati, OH (1976).

62. Dupuis, H.: Mechanische Schwingungen, sowie: Messung und Bewertung von Schwingungen und Stössen. In *Ergonomie,* Band 2, pp. 211−236, ed. by H. Schmidtke. Carl Hanser Verlag, Munich (1974).

63. Durnin, J. V. G. A. and Passmore, R.: *Energy, Work and Leisure.* Heinemann Educational, London (1967).

64. Egli, R., Grandjean, E. and Turrian, H.: Arbeitsphysiologische Untersuchungen an Hackgeräten. *Arbeitsphysiologie,* **15,** 231−234 (1943).

65. Eastman, M. C. and Kamon, E.: Posture and subjective evaluation at flat and slanted desks. *Human Factors,* **18,** 15−26 (1976).

66. Elias, R. and Cail, F.: Exigences visuelles et fatigue dans deux types de tâches infomatisées. *Le Travail Humain,* **46,** 81−92 (1983).

67. Ellis, D. S.: Speed of manipulative performance as a function of worksurface height. *Journal of Applied Psychology,* **35,** 289−296 (1951).

68. Engel, F. L.: Information selection from visual display units. In *Ergonomic Aspects of Visual Display Terminals,* ed. by E. Grandjean and E. Vigliani. Taylor & Francis, London (1980).

69. Eysel, U. T. and Burandt, U.: Fluorescent tube light evokes

flicker responses in visual neurons. *Vision Research*, 24, 943−948 (1984).

70. Fanger, P. O.: *Thermal Comfort*. McGraw-Hill, New York (1972). *Thermal Comfort−Analysis and Application in Environmental Engineering*. Danish Technical Press, Copenhagen (1970).

71. Fletcher and Munson, W. A.: Loudness, its definition, measurement and calculation. *Journal of the Acoustical Society of America*, 5, 82−108 (1933).

72. Frankenhäuser M.: Man in Technological Society: Stress, Adaptation and Tolerance Limits. Report from the Psychological Laboratories, University of Stockholm, Suppl. 26 (1974).

73. Frankenhäuser, M., Nordheden, B., Myrsten, A. L. and Post, B.: Psychophysiological reactions to understimulation and overstimulation. *Acta Psychologica*, 35, 298−308 (1971).

74. Freivalds, A., Chaffin, D. B., Garg, A. and Lee, K. S.: A dynamic biochemical evaluation of lifting maximum acceptable loads. *Journal of Biomechanics*, 17, 251−262 (1984).

75. Friedmann, G: *Grenzen der Arbeitsteilung*. Europäische Verlagsanstalt, Frankfurt (1959).

76. Fröberg, J., Karlsson, C. G. and Levi, L.: Shiftwork, a study of catecholamine excretion, selfratings and attitudes. *Studia Laboris et Salutis*, 11, 10−20 (1972).

77. Fry, G. A. and King, V. M.: The pupillary response and discomfort glare. *Journal of the IES*, 307−324 (1975).

78. General Electric Co.: See Better, Work Better. Lamp Division, Bulletin No. 1, Cleveland (1953).

79. Geyer, B. H. and Johnson, C. W.: Memory in man and machines. *General Electric Review*, 60, 29−33 (1957).

80. Gierer, R., Martin, E., Baschera, P. and Grandjean, E.: Ein neues Gerät zur Bestimmung der Flimmerverschmelzungsfrequenz. *Zeitschrift für Arbeitswissenschaft*, 35, 45−47 (1981)

81. Graf, O.: Studien über Fliessarbeitsprobleme an einer praxisnahen Experimentieranlage. Forschungsbericht d. Wirtschafts- und Verkehrsministerium Nordrhein-Westfalen, Nos 114 and 115. Westdeutscher Verlag, Cologne (1954).

82. Grandjean, E.: Physiologische Untersuchungen über die nervöse Ermüdung bei Telephonistinnen und Büroangestellten, *Internationale Zeitschrift Angewandte Physiologie* 17, 400−418 (1959)

83. Grandjean, E.: Fatigue. Yant Memorial Lecture 1970 *American Industrial Hygiene Association Journal*, 31, 401−411 (1970).

84. Grandjean, E.: *Ergonomics of the Home*. Taylor & Francis, London (1973).

85. Grandjean, E.: *Ergonomics in Computerized Offices*. Taylor & Francis, London (1987).

86. Grandjean, E. and Burandt, H. U.: Das Sitzverhalten von Büroangestellten. *Industrielle Organisation*, 31, 243−250 (1962).

87. Grandjean, E., Egli, R., Rhiner, A. and Steinlin. H.: Der menschliche Energieverbrauch der gebräuchlichsten Waldsägen. *Helvetica Physiologie et Pharmacologie Acta*, 10, 342−348 (1952).

88. Grandjean, E., Horisberger, B., Havas, L, and Abt, K.: Arbeits-

physiologische Untersuchungen mit verschiedenen Beleuchtungssystemen an einer Feinarbeit. *Industrielle Organisation*, **28**, 231–239 (1959).

89. Grandjean, E., Streit, K. and Perret, E.: Arbeitsphysiologische Untersuchungen über die Ermüdung bei Tätigkeiten, mit geringen und hohen Anforderungen an die Aufmerksamkeit. *Arbeitsmedizin, Sozialmedizin und Arbeitshygiene*, **1**, 2–11 (1966).

90. Grandjean, E., Böni, A. and Kretschmar, H.: Entwicklung eines Ruhesesselprofils für gesunde und rückenkranke Menschen. *Wohnungsmedizin*, **5**, 51–56 (1967).

91. Grandjean, E., Wotzka, G. Schaad, R. and Gilgen, A.: Fatigue and stress in air traffic controllers, *Ergonomics*, **14**, 159–165 (1971).

92. Grandjean, E., Hünting, W., Wotzka, G. and Schärer, R.: An ergonomic investigation of multipurpose chairs. *Human Factors*, **15**, 247–255 (1973).

93. Grandjean, E., Nakaseko, M., Hünting, W. and Läubli, T.: Ergonomische Untersuchungen zur Entwicklung einer neuen Tastatur für Büromaschinen. *Zeitschrift der Arbeitswissenschaft* , **35**, 221–226 (1981).

94. Grandjean, E., Hünting, W. and Pidermann M.: VDT workstation design: preferred settings and their effects. *Human Factors*, **25**, 161–175 (1983).

95. Grether, W. F.: Vibration and human performance. *Human Factors*, **13**, 203–216 (1971).

96. Griffin, M. J.: Whole-body vibration levels affecting visual acuity. In *Human Reaction to Vibration*. Proceedings of the Conference, Salford University (1973).

97. Guignard, J. C. and King, P. F.: *Aeromedical Aspects of Vibration and Noise*. NATO Advisory Group for Aerospace Research and Development, A D 754631 (1972).

98. Guth, S. K.: Light and comfort. *Industrial Medicine and Surgery*, **27**, 570–574 (1958).

99. Hagberg, M.: Arbetsrelaterade besvär i halsrygg och skuldra. Swedish Work Environment Fund, Report, Stockholm (1982).

100. Haggard, H. W. and Greenberg, L. A.: *Diet and Physical Efficiency*. Yale University Press, New Haven, CT (1935).

101. Haider, M.: Experimentelle Untersuchungen über Daueraufmerksamkeit und cerebrale Vigilanz bei einförmigen Tätigkeiten. *Zeitschrift Experimentale und Angewandte Psychologie*, **10**, 1–18 (1963).

102. Harris, W., Mackie, R. R. *et al.*, A Study of the Relationship among Fatigue, Hours of Service and Safety of Operations of Truck and Bus Drivers. Human Factor Research Inc. Santa Barbara, Goleta, CA 93017, Report No. 1727–2 (1972).

103. Harrison, R. V.: Person-environment fit and job stress. In *Stress at Work*, ed. by C. Cooper and R. Payne. John Wiley, New York (1978).

104. Hashimoto, K.: Physiological features of monotony manifested under high speed driving situations. In *Proceedings of the 16th International Congress of Occupational Health*, Tokyo, 85–88 Railway Labour Science Institute, Japan National Railways. (1969).

105. Haubner, P. and Kokoschka, S.: Visual Display Units—Characteristics of Performance. International Commission on Illumination (CIE), 20th Session in Amsterdam, 52 Bd. Malesherbes 75008 Paris, France (1983).

106. Hawel, W.: Untersuchungen eines Bezugssystems für die psychologische Schallbewertung. *Arbeitswissenschaft*, **6**, 123—127 (1967).

107. van der Heiden, G. and Krüger, H.: Evaluation of Ergonomic Features of the Computer Vision Instaview Graphics Terminal. Report of the Dept. of Ergonomics, Swiss Federal Institute of Technology, 8092 Zürich (1984).

108. van der Heiden, G., Bräuninger, U. and Grandjean, E.: Ergonomic studies on computer aided design. In *Ergonomics and Health in Modern Offices*, ed. by E. Grandjean. Taylor & Francis, London (1984).

109. den Hertog, F. J. and Kerkhoff, W. H. C.: Vom Fliessband zur selbständigen Gruppe. *Industrielle Organisation*, **43**, 21—24 (1974).

110. Hertzberg, H. T. E., Daniels, G. S. and Churchill, E.: Anthropometry of Flying Personnel—1950. USAF Wright Air Development Center, Tech. Report, 52—321 (1954).

111. Hess, W. R.: *Die funktionelle Organisation des vegetativen Nervensystems*. Benno Schwabe, Basel, (1948).

112. Hettinger, T.: Muskelkraft bei Männern und Frauen. *Zentralblatt Arbeit und Wissenschaft*, **14**, 79—84 (1960).

113. Hettinger, T.: *Angewandte Ergonomie*. Bartmann-Verlag, Frechen, FR Germany (1970).

114. Hettinger, T. and Müller, E. A.: Der Einfluss des Schuhgewichtes auf den Energieumsatz beim Gehen und Lastentragen. *Arbeitsphysiologie*, **15**, 33—40 (1953).

115. Hettinger, T., Kaminsky, G. and Schmale, H.: *Ergonomie am Arbeitsplatz*. Friedrich Kiehl Verlag, Ludwigshafen (1976).

116. Hilgendorf, L.: Information input and response time. *Ergonomics*, **9**, 31—37 (1966).

117. Hill, J. H. and Chernikoff, R.: Altimeter Display Evaluation: Final Report USN, NEL Report 6242 Jan 26 (1965).

118. Hill, S. G. and Kroemer, K.: Preferred declination of the line of sight angle. *Ergonomics*, **29**, 1129—1134 (1986).

119. Hort, E.: A new concept in chair design. Proceedings of Ergodesign 84, *Behaviour and Information Technology*, **3**, 359—362 (1984).

120. Houghten, F. C. and Yaglou, C. P.: Determining lines of equal comfort. *ASHVE Transactions*, **29**, 163—171 (1923).

121. Hoyos, C.: Kompatibilität in *Ergonomie Band 2*, ed. by H. Schmidtke. Carl Hanser Verlag, Munich (1974).

122. Hünting, W. and Grandjean, E.: Sitzverhalten und subjektives Wohlbefinden auf schwenkbaren und fixierten Formsitzen. *Zeitschrift der Arbeitswissenschaft*, **30**, 161—164 (1976).

123. Hünting, W., Nemecek, J. and Grandjean, E.: Die physische Belastung von Arbeitern an der Gesenkschmiede—eine Fallstudie. *Sozial- und Präventivmedizin*, **19**, 275—278 (1974).

124. Hünting, W., Läubli, T. and Grandjean, E.: Postural and visual loads at VDT workplaces, Part 1: Constrained postures. *Ergonomics*, **24**, 917—931 (1981).

125. Hünting W., Nakaseko, M., Gierer, R. and Grandjean, E.: Ergonomische Gestaltung von alphanumerischen Tastaturen. *Sozial- und Präventivmedizin*, **27**, 251−252 (1982).

126. IBM: Human factors of workstations with visual displays. IBM Human Factors Center, Dept. P15, Bldg 078. 5600 Cottle Rd, San Jose, CA 95193 (1984).

127. IES: *IES Lighting Handbook*, 5th ed. Illuminating Engineering Society, New York (1972).

128. ISO (International Organization for Standardization): TC 43, Assessment of Noise-Exposure during Work for Hearing Conversation Purposes. Geneva (1971).

129. ISO (International Organization for Standardization): 2631: Guide for the Evaluation of Human Exposure to Whole-Body Vibration. Geneva (1974).

130. Jasper, H.: Quoted from W. F. Ganong: *Lehrbuch der Medizinischen Physiologie*, Deutsche Ausgabe, Springer Verlag, Berlin. (1974)

131. Jenkins, W. O.: The tactual discrimination of shapes for coding aircraft type controls. In Psychological Research on Equipment Design, ed. by P. M. Fitts. Army Air Force, Aviation Psychology Program, Report 19 (1947).

132. Jerison, H. J.: Effects of noise on human performance. *Journal of Applied Psychology*, **43**, 96−101 (1959).

133. Jerison, H. J. and Pickett, R. M.: Vigilance: the importance of the elicited observing rate. *Science*, **143**, 970−971 (1964).

134. Johansson, G.: In: *Human Aspects in Office Automation*, Elsevier Series in Office Automation, Vol. 1, ed. by B. G. F. Cohen. Elsevier, New York (1984).

135. Johansson, G. and Aronsson, G.: Stress Reactions in Computerized Administrative Work. Report from the Department of Psychology, University of Stockholm, Suppl. 50, November (1980).

136. Johansson, G., Aronsson, G., and Lindström, B. O.: Social, Psychological and Neuroendocrine Stress Reactions in Highly Mechanized Work. Report from the Psychological Laboratories, University of Stockholm, No. 488 (1976).

137. Jürgens, H. W.: Körpermasse. in *Ergonomie*, Band 1, ed. by H. Schmidtke. Carl Hanser Verlag, Munich (1973).

138. Kahn, H.: *Toward the Year 2000*. American Academy of Arts and Sciences (1967).

139. Kalsbeck, J. W. H.: Sinus arrhythmia and the dual task method in measuring mental load. In *Measurement of Man at Work*, ed. by W. T. Singleton, J. G. Fox and D. Whitfield. Taylor & Francis, London (1971)

140. Karrasch, K. and Müller, E. A.: Das Verhalten der Pulsfrequenz in der Erholungsperiode nach körperlicher Arbeit. *Arbeitsphysiologie*, **14**, 369−382 (1951).

141. Keegan, J. J.: Alterations of the lumbar curve related to posture and seating. *Journal of Bone and Joint Surgery*, **35**, 567−589 (1953).

142. Kerr, W. A.: Worker attitudes toward scheduling of industrial music. *Journal of Applied Psychology*, **30**, 575−578 (1946).

143. Klockenberg, E. A.: *Rationalisierung der Schreibmaschine und ihrer Bedienung*. Springer, Berlin (1926).

144. Knauth, P. and Rutenfranz, J.: Das Verhalten der Körpertemperatur in verschiedenen Schichtsystemen bei experimenteller Schichtarbeit. *Zeitschrift für Arbeitswissenschaft*, **31**, 18−21 (1977).

145. Koch, K. W., Jennings, B. H. and Humphreys, C. H.: Is humidity important in the temperature comfort range? *ASHRAE Transactions* **66**, 63−68 (1960).

146. Krämer, J. *Biomechanische Veränderungen im lumbalen Bewegungssegment*, Hippokrates, Stuttgart (1973).

147. Kroemer, K. H. E.: Uber den Einfluss der räumlichen Lage von Tastenfeldern auf die Leistung an Schreibmaschinen, *Internationale Zeitschrift Physiologie*, **20**, 453−464 (1965).

148. Kroemer, K. H. E.: Was man von Schaltern, Kurbeln und Pedalen wissen muss. Sonderheft der *REFA-Nachrichten*, Verband für Arbeitsstudien, REFA e.V., Darmstadt (1967).

149. Kroemer, K. H. E.: Foot operation of controls. *Ergonomics*, **14**, 333−339 (1971).

150. Kroemer, K. H. E.: Human engineering—the keyboard. *Human Factors*, **14**, 51−63 (1972).

151. Kromodihardjo, S. and Mital, A.: Kinetic analysis of manual lifting activities: Part 1, Development of a three-dimensional computer model. *International Journal of Industrial Ergonomics*, **1**, 77−90 (1986).

152. Krueger, H.: Zur Ergonomie von Balans-Sitzelementen im Hinblick auf ihre Verwendbarkeit als reguläre Arbeitsstühle. Report of the Dept. of Ergonomics, Swiss Federal Institute of Technology, 8092 Zürich (1984).

153. Krueger, H. and Hessen, J.: Objective kontinuierliche Messung der Refraktion des Auges. *Biomedizinische Technik*, **27**, 142−147 (1982).

154. Krueger, H. and Müller-Limmroth, W.: Arbeiten mit dem Bildschirm—aber richtig! Bayerisches Staatsministerium für Arbeit und Sozialordnung, Winzererstr 9, 8000 Munich 40 (1979).

155. Kryter, K. D: Damage risk criterion and contours based on permanent and temporary hearing loss data. *American Industrial Hygiene Association Journal*, **26**, 34−44 (1965).

156. Läubli, T.: Das arbeitsbedingte cervicobrachiale Überlastungssyndrom, Thesis, Medical Faculty, University of Zurich (1981).

157. Läubli T. and Grandjean, E. The magic of control groups in VDT field studies. In *Ergonomics and Health in Modern Offices*, ed. by E. Grandjean. Taylor & Francis, London (1984).

158. Läubli, T., Hünting, W. and Grandjean, E.: Postural and visual loads at VDT workplaces, Part 2: Lighting conditions and visual impairments. *Ergonomics*, **24**, 933−944 (1981).

159. Läubli, T., Senn, E., Fasser, W., Mion, H., Carlo, T. and Zeier, H.: Klinische Befunde und subjektive Klagen über Beschwerden im Bewegungsapparat. *Sozial- und Präventivmedizin*, **31**, (2) (1986).

160. Lazarus, R. S.: Cognitive and coping processes in emotion. In *Stress and Coping*, ed. by A. Monat and R. S. Lazarus. Columbia University Press, New York (1977).

161. Lecret, F.: La Fatigue du Conducteur. Cahier d'etude de l'Or-

ganisme National de Sécurité Routière (ONSER) Bull. No. 38, Paris (1976).

162. Lehmann, G.: *Praktische Arbeitsphysiolgie*, 2nd edition. Thieme Verlag, Stuttgart (1962).

163. Lehmann, G. and Stier, F.: Mensch und Gerät. *Handbuch der gesamten Arbeitsmedizin*. Vol. 1, pp. 718–788. Urban und Schwarzenberg, Berlin (1961).

164. Leithead, C. S. and Lind, A. R.: *Heat Stress and Heat Disorders*. Cassell, London (1964).

165. Leplat, J.: Attention et incertitude dans les travaux de surveillance et d'inspection. *Sciences du Comportement*, No. 6, Dunod, Paris (1968).

166. Levi, L.: *Emotions—Their Parameters and Measurement*, Raven Press, New York (1975).

167. Lille, F.: Le Sommeil de jour d'un groupe de travailleurs de nuit. *Le Travail Humain*, **30**, 85–97 (1967).

168. Lind, A. R. and McNicol, G. W.: Cardiovascular responses to holding and carrying weight by hand and by shoulder harness. *Journal of Applied Physiology*, **25**, 261–267 (1968).

169. Luckiesh, H. and Moss, F. K.: *The Science of Seeing*. Van Nostrand, New York (1937).

170. Lundervold, A.: Electromyographic investigations of position and manner of working in typewriting. *Acta Physiologica Scandinavia*, **84**, 171–183 (1951).

171. Lundervold, A.: Electromyographic investigations during typewriting. *Ergonomics*, **1**, 226–233 (1958).

172. Mackay, C., Cox, T., Burrows, G. and Lazzerini, T.: An inventory for the measurement of selfreported stress and arousal. *British Journal of Social and Clinical Psychology*, **17**, 283–284 (1978).

173. Mackworth, J. F.: *Vigilance and Habituation*. Penguin Books, Harmondsworth (1969).

174. Mackworth, N. H.: Research on the Measurement of Human Performance. HMSO, London (1950).

175. Maeda, K., Horiguchi, S. and Hosokawa, M.: History of the studies on occupational cervicobrachial disorder in Japan and remaining problems, *Journal of Human Ergology*, **11**, 17–29 (1982).

176. Malhotra, M. S. and Sengupta, J.: Carrying of school bags by children. *Ergonomics*, **8**, 55–60 (1965).

177. Mandal, A. C.: What is the correct height of furniture? In: *Ergonomics and Health in Modern Offices*, ed. by E. Grandjean. Taylor & Francis, London (1984).

178 Maric, D.: *L'Aménagement du Temps de Travail*. Bureau International du Travail, 1211 Geneva (1977).

179. Martin, E., and Weber, A.: Wirkungen eintönig-repetitiver Tätigkeiten auf das subjektive Befinden und die Flimmerverschmelzungsfrequenz. *Zeitschrift für Arbeitswissenschaft*, **30**, 183–187 (1976).

180. McCormick, E. J. and Sanders, M., *Human Factors in Engineering*. 5th edition. McGraw-Hill, New York (1987).

181. McConnell, W. J. and Spiegelman, M.: Reactions of 745 clerks to summer air conditioning. *Heating, Piping, Air Conditioning*, **12**, 317–322 (1940).

182. McConnell, W. J. and Yaglou, C. P.: Work tests conducted in

atmospheres of high temperatures and various humidities in still and moving air. *Journal of the American Society of Heating and Ventilation Engineers.* **31**, 217−221 (1925).

183. McFarland, R. A.: *Human Factors in Air Transport Design.* McGraw-Hill, New York (1946).

184. McGrath, J. E.: Stress and behaviour in organisations. In *Handbook of Industrial and Organisational Psychology*, ed. by M. Dunnette. Rand McNally, Chicago (1976).

185. Ministry of Defence, London: Human Factors for Designers of Equipment. Part 3: Body Strength and Stamina. MoD 00−25, London (1984).

186. Mital, A., and Kromodihardjo, S.: Kinetic analysis of manual lifting activities: Part 2, Biochemical analysis of task variables. *International Journal of Industrial Ergonomics*, **1**, 91−101 (1986).

187. Monod, H.: La dépense energétique chez l'homme. In *Physiologie du Travail*, ed. by J. Scherrer. Masson, Paris (1967).

188. Morgan, C. T., Chapanis, A., Cook, J. S. and Lund, M. W.: *Human Engineering Guide to Equipment Design.* McGraw-Hill New York (1963).

189. Mott, P. E., Mann, C., McLoughlin and Warwick, P.: *Shiftwork: The Social, Psychological and Physical Consequences.* University of Michigan Press, Ann Arbor (1965).

190. Müller, E. A.: Die physische Ermüdung. In *Handbuch der gesamten Arbeitsmedizin*, Band 1. Urban & Schwarzenberg, Berlin (1961).

191. Müller, E. A. and Spitzer, H.: *Arbeit recht verstanden.* Oldenbourg Verlag, Munich (1952).

192. Müller-Limmroth, W.: Sinnesorgane. In: *Ergonomie* Vol. 1 ed. by H. Schmidtke. Carl Hanser Verlag, Munich (1973).

193. Murrell, K. F. H.: *Ergonomics; Man in his Working Environment.* Chapman and Hall, London (1965).

194. Murrell, K. F. H.: *Human Performance in Industry.* Reinhold, New York (1965).

195. Nachemson, A.: Lumbal intradiscal pressure. Results from in vitro and in vivo experiments with some clinical implications. *7. Wissenchafticher Konferenz. Deutscher Naturforscher and Aerzte.* Springer, Berlin, (1974).

196. Nachemson, A. and Elfström, G.: Intravital dynamic pressure measurements in lumbar discs. *Scandinavian Journal of Rehabilitation Medicine*, Suppl. 1. (1970).

197. Nakaseko, M., Grandjean E., Hünting, W. and Gierer, R.: Studies on ergonomically designed alphanumeric keyboards, *Human Factors*, **27**, 175−187 (1985).

198. Nemecek, J. and Grandjean, E.: Das Grossraumbüro in arbeitsphysiologischer Sicht. *Industrielle Organisation*, **40**, 233−243 (1971).

199. Nemecek, J. and Grandjean, E.: Etude ergonomique d'un travail pénible dans l'industrie textile. *Le Travail Humain*, **38**, 167−174 (1975).

200. Neumann, J. and Timpe, K. P.: *Arbeitsgestaltung. Psychophysiologische Probleme bei Ueberwachungs- und Steuerungstätigkeiten.* VEB Deutscher Verlag der Wissenschaften, Berlin (1970).

201. Nevins, R. G., Rohles, F. H., Springer, W. and Feyerherm, A.

M.: A temperature-humidity chart of thermal comfort of seated persons. *ASHRAE Journal*, **8**, 55−61 (1966).

202. NIOSH: *Work Practices Guide for Manual Lifting*. National Institute for Occupational Safety and Health, Cincinnati, OH (1981).

203. Nishiyama, K., Nakaseko, M. and Uehata, T.: Health aspects of VDT operators in the newspaper industry. In *Ergonomics and Health in Modern Offices*, ed. by E. Grandjean. Taylor & Francis, London (1984).

204. Nitsch, J.: Theorie und Skalierung der Ermüdung. Dissertation, Cologne (1970).

205. Northrup, H. R: *Hours of Work*. Harper and Row, New York (1965).

206. O'Hanlon, J. F.: Heart Rate Variability: A New Index of Drivers' Alertness/Fatigue. Human Factors Research Inc. Santa Barbara, Goleta, CA. Report No., 1812−1 (1971).

207. O'Hanlon, J. F., Royal, J. W. and Beatty, J.: EEG Theta Regulation and Radar Monitoring Performance in a Controlled Field Experiment. Human Factors Research Inc., Santa Barbara, CA. Report No. 1738−F (1975).

208. Pearson, R. G. and Byars, G. E.: The Development and Validation of a Checklist for Measuring Subjective Fatigue. Report 56−115. School of Aviation Medicine, USAF, Randolph AFB, TX. (1956).

209. Pheasant, S.: *Bodyspace: Anthropometry, Ergonomics and Design*. Taylor & Francis, London (1986).

210. Pierce, J. R. and Karlin, J. E: Reading rates and the information rate of a human channel. *Bell Telephone Journal*, **36**, 497−516 (1957).

211. Proceedings of the International Conference on Enhancing the Quality of Working Life. Arden House, Harriman, New York (Sept. 1972).

212. Prokop, O. and Prokop, L.: Ermüdung und Einschlafen am Steuer. *Deutsche Zeitschrift für gerichtliche Medizin*, **44**, 343−350 (1955).

213. Rey, P. and Rey, J. P.: Effect of an intermittent light stimulation on the critical fusion frequency. *Ergonomics*, **8**, 173−180 (1965).

214. Robinson, D. W. and Dadson, R. S.: Threshold of hearing and equal-loudness relations for pure tones and the loudness function. *Journal of the Acoustical Society of America*, **29**, 1284−1288 (1957).

215. Robinson, G. and Gerking, S. D.: The thermal balance of men working in severe heat. *American Journal of Physiology*, **149**, 102−108 (1947).

216. Rohmert, W.: Die Grundlagen der Beurteilung statischer Arbeit. Forschungsberichte des Landes Nordrhein-Westfalen No. 938. Westdeutscher Verlag, Cologne (1960).

217. Rohmert, W.: Statische Haltearbeit des Menschen. Special issue of *REFA-Nachrichten* (1960).

218. Rohmert, W.: Maximalkräfte von Männern im Bewegungsraum der Arme und Beine. Forschungsberichte des Landes Nordrhein-Westfalen No. 1616. Westdeutscher Verlag, Cologne (1966).

219. Rohmert, W. and Hettinger, T.: Ergebnisse achtstündiger

Untersuchungen am Kurbel- und Fahrradergometer. Quoted in (117).

220. Rohmert, W. and Jenik, P.: *Maximalkräfte von Frauen im Bewegungsraum der Arme und Beine.* In the series Arbeitswissenschaft und Praxis. Beuth-Vertrieb, Berlin (1972).

221. Rohmert, W., Rutenfranz, J. and Ulich, E.: Das Anlernen sensumotorischer Fertigkeiten. Europäische Verlagsanstalt, Frankfurt (1971).

222. Rosemeyer, B.: Elektromyographische Untersuchungen der Rücken und Schultermuskulatur im Stehen und Sitzen unter Berücksichtigung der Haltung des Autofahrers. *Archiv orthopaedische Unfall-Chirurgie,* **69**, 59—70 (1971).

223. Rosenkranz, R.: Die Viertagewoche und die kritische Untersuchung des lst-Zustandes. *Rationelles Büro,* **22**, 7 (1971).

224. Rupp, B. A.: Visual display standards: a review of issues. *Proceedings of the Society for Information Display,* **22**, 63—72 (1981).

225. Rutenfranz, J. and Knauth, P.: Rhythmusphysiologie und Schichtarbeit. In *Schicht- und Nachtarbeit.* Institut für Gesellschaftspolitik; Sensenverlag, Vienna (1976).

226. Ryan, A. H. and Warner, M.: The effect of automobile driving on the reactions of the driver. *American Journal of Physiology,* **48**, 403—409 (1936).

227. Saito, H., Kishida, K., Endo, Y. and Saito, M.: Studies on bottle inspection task. *Journal of Science of Labour,* **48**, 475—525 (1972).

228. Salvendy, G.: Research issues in the ergonomics, behavioural, organizational and management aspects of office automation. In *Human Aspects in Office Automation,* ed. by B. G. F. Cohen. Elsevier, New York (1984).

229. Sauter, S. L.: Predictors of strain in VDT users and traditional office workers. In: *Ergonomics and Health in Modern Offices,* ed. by E. Grandjean. Taylor & Francis, London (1984).

230. Sauter, S. L., Gottlieb, M. S., Jones, K. C., Dodson, V. N. and Rohrer, K. M.: Job and Health implications of VDT use: Initial results of the Wisconsin-NIOSH study. *Communications of the ACM,* **26**, 284—294 (1983).

231. Scherrer, J.: Physiologie Musculaire. In *Physiologie du Travail,* ed. by J. Scherrer. Masson, Paris (1967).

232. Schmidtke, H.: Wachsamkeitsprobleme. In H. Schmidtke, *Ergonomie,* Band 1, ed. by H. Schmidtke. Carl Hanser Verlag, Munich (1973).

233. Schmidtke, H.: Bedienungs- und Steuerarmaturen. In *Ergonomie,* Band 2, ed. by H. Schmidtke, Carl Hanser Verlag, Munich (1974).

234. Schmidtke, H. and Stier, F.: Der Aufbau komplexer Bewegungsabläufe aus Elementarbewegungen. Forschungsbericht des Landes Nordrhein-Westfalen, No. 822, Westdeutscher Verlag, Cologne (1960).

235. Schoberth, H.: *Sitzhaltung — Sitzschaden — Sitzmöbel.* Springer, Berlin (1962).

236. Scholz, H.: Die physische Arbeitsbelastung der Giessereiarbeiter. Forschungsbericht des Landes Nordrhein-Westfalen No. 1185, Westdeutscher Verlag, Cologne (1963).

237. Selye, H.: *The Stress of Life.* McGraw-Hill, New York (1956).

238. Shackel, B.: *Applied Ergonomics Handbook*. Reprints from *Applied Ergonomics* Vols 1 and 2. IPC Science and Technology Press, Guildford (1974).

239. Shahnavaz, H.: Lighting conditions and workplace dimensions of VDU operators. *Ergonomics*, **25**, 1165−1173 (1982).

240. Shannon, C. E. and Weaver, W.: *The Mathematical Theory of Communication*. University of Illinois Press, Urbana (1949).

241. Shute S. J., and Starr S. J.: Effects of adjustable furniture on VDT users. *Human Factors*, **26**, 157−170 (1984).

242. Simons, A. K., Radke, A. O. and Oswald, W. C.: A Study of Truck Ride Characteristics in Military Vehicles. Bostrom Research Laboratories, Milwaukee, Report 118 (1956).

243. Singleton, W. T.: *Introduction à l'Ergonomie*. Organisation Mondiale de la Santé, Geneva (1974).

244. Sleight, R, B.: The effect of instrument dial shape on legibility. *Journal of Applied Psychology*, **32**, 170−188 (1948).

245. Smith, M. J., Stammerjohn, L. W., Cohen, B. G. F., and Lalich, N. R.: Job stress in video display operations. In *Ergonomic Aspects of Visual Display Terminals*, ed. by E. Grandjean and E. Vigliani. Taylor & Francis, London (1980).

246. Smith, M. J., Cohen, B. C. F., Stammerjohn, L. W. and Happ, A.: An investigation of health complaints and job stress in video display operations. *Human Factors*, **23**, 387−399 (1981).

247. Snook, S. H.: The design of manual handling tasks. *Ergonomics*, **21**, 963−985 (1978).

248. Steinbuch, K.: Information processing in man. Quoted in McCormick and Sanders (1984).

249. Stier, F. and Meyer, H. O.: Physiologische Grundlagen der Arbeitsgestaltung. Verband für Arbeitsstudien (REFA) Darmstadt.

250. Swensson, A. (ed.): *Night and Shiftwork*. Proceedings of an International Symposium, Nat. Institute of Occupational Health, Stockholm, Sweden (1969), and a second symposium: Studia Laboris et Salutis, Rep. No. 11, Stockholm (1972).

251. Swink, J. R.: Intersensory comparisons of reaction time using an electropulse tactile stimulus. *Human Factors*, **8**, 143−145 (1966).

252. Taylor, C. L.: The biomechanics of the normal and of the amputated upper extremity. In *Human Limbs and Their Substitutes*, pp. 169−221, ed. by Klopsteg and Wilson. McGraw Hill, New York, S. 169−221 (1954).

253. Teeple, J. B.: Work of carrying loads. *Journal of Perceptual and Motor Skills*, **7**, 60−68 (1957).

254. Thiberg, S.: Anatomy for Planners, Parts I−IV. Statens Institut für Byggnadsforskning (1965−1970).

255. Thiis-Evenson, E.: Shiftwork and health. *Industrial Medicine*, **27**, 493−497 (1958).

256. Tichauer, E. R.: Potential of biomechanics for solving specific hazard problems. *Conference of the American Society of Safety Engineers*, pp. 149−187, Park Ridge, Il. (1968).

257. Tichauer, E. R.: Occupational Biomechanics. Rehabilitation Monograph No. 51, New York University, Center for Safety (1975).

258. Tichauer, E. R.: Biomechanics sustains occupational safety and health. *Industrial Engineering*, **27**, 46−56 (1976).

259. Timmers, H.: An effect of contrast on legibility of printed text. *I.P.O. Annual Progress Report*, No. 13, 64−67 (1978).

260. Troup, J. D. G. and Edwards, F. C.: *Manual Handling — A Review Paper*. HMSO, London (1985).

261. US Dept. of Health, Education and Welfare: *Weight, Height and Selected Body Dimensions of Adults: United States 1960−62*. Vital and Health Statistics, Series 11, no. 8, Government Printing Office, Washington (1966).

262. VDI Richtlinie 2058: *Beurteilung von Arbeitslärm am Arbeitsplatz hinsichtlich Gehörschäden*. Blatt 2. VDI Verlag, Düsseldorf (1970).

263. Vernon, M. H.: *Industrial Fatigue and Efficiency*. Dutton, New York (1921).

264. Verwaltungs-Berufsgenossenschaft: *Sicherheitsregeln für Bildschirm-Arbeitsplätze im Bürobereich*. Überseering 8, 2000 Hamburg (1981).

265. Wakim, K. G. Gersten, J. W., Elkins, E . C. and Martin, G. M.: Objective recording of muscle strength. *Archives of Physical Medicine*, **31**, 90−99 (1950).

266. Ward, W. D.: Noise-induced hearing damage. In *Otolaryngology* Vol. 2, ed. by M. M. Paparella and D. A. Shumrick. Saunders, Philadelphia (1973).

267. Wargo, M. J.: Human operator response speed, frequency and flexibility: a review and analysis. *Human Factors*, **9**, 221−238 (1967).

268. Warrick, M. J., Kibler, A. W. and Topmiller, D. A.: Response time to unexpected stimuli. *Human Factors*, **9**, 81−86 (1965).

269. Weber, A.: Irritating and annoying effects of passive smoking. In *Ergonomics and Health in Modern Offices*, ed. by E. Grandjean. Taylor & Francis, London (1984).

270. Weber A., Jermini, C. and Grandjean, E.: Beziehung zwischen objektiven und subjektiven Messmethoden bei experimentell erzeugter Ermüdung. *Zeitschrift für Präventivmedizin*, **18**, 279−283 (1973).

271. van Wely, P.: Design and disease. *Applied Ergonomics*, **1**, 262−269 (1970).

272. Wenzel, H. G.: Möglichkeiten und Probleme der Beurteilung von Hitzebelastungen des Menschen. *Arbeitswissenschaft*, **3**, 73−83 (1964).

273. Wiebelitz, R. and Schmitz, H.: Flimmerndes Licht und die menschliche Augenpupille. *Zeitschrift der Arbeitswissenchaft*, **37**, 163−168 (1983).

274. Wirths, W.: Ist eine Zwischenverpflegung während der Arbeitszeit ernährungsphysiologisch notwendig? In *Ernährungspädagogisches Colloquium*, Mühlenstelle, Bonn, Bericht 10 (1976).

275. Wisner, A.: Audition et Bruits. In *Physiologie du Travail*, Vol. 2, ed. by J. Scherrer. Masson, Paris (1967).

276. Wisner, A.: Effets des vibrations sur l'homme. In *Physiologie du Travail*, Vol. 2. ed. by J. Scherrer. Masson, Paris (1967).

277. Woodson, W. E. and Conover, D. W.: *Human Engineering Guide for Equipment Designers*. University of California Press, Berkeley (1964).

278. Wyatt, S. and Marriott, R.: A study of attitudes to factory work. *MRC Special Report Series*, **292**, London (1956).

279. Wyndham, C. H. *et al.*: Examination of heat stress indices. *Archives of Industrial Hygiene and Occupational Medicine*, **7**, 221−233 (1953).
280. Yaglou, C. P., Riley, E. C. and Coggins, D. I.: Ventilation requirements, *ASHVE Transactions*, **42**, 133−158 (1936).
281. Yaglou, C. P., Riley, E. C. and Coggins, D.I.: *Ventilation Requirements and the Science of Clothing*. Saunders, Philadelphia (1949).
282. Yamaguchi, Y., Umezawa, F. and Jshinada, Y.: Sitting posture: an electromyographic study on healthy and notalgic people. *Journal of the Japanese Orthopedics Association*, **46**, 51−56 (1972).
283. Yllö, A.: The biotechnology of card-punching. *Ergonomics*, **5**, 75−79 (1962).
284. Zeier, H. and Bättig, K.: Psychovegetative Belastung und Aufmerksamkeitsspannung von Fahrzeuglenkern auf Autobahnabschnitten mit und ohne Geschwindigkeitsbegrenzung. *Zeitschrift für Verkehrssicherheit*, **23**, 1 (1977).
285. Zipp, P., Haider, E., Halpern, N. and Rohmert, W.: Keyboard design through physiological strain measurements. *Applied Ergonomics*, **14**, 117−122 (1983).

Index